Introduction

Coming to Terms with Maths is a **valuable resource** that is designed to help students understand the mathematical terms, symbols and formulae in textbooks for the **Junior Certificate** and **Leaving Certificate** programmes.

Coming to Terms with Maths features:

- **A–Z definitions** of many **mathematical terms** students encounter from First Year right up to Leaving Certificate Higher Level
- clearly labelled explanatory diagrams
- explanations of **100 key formulae**
- a list of **key mathematical symbols** and their meanings
- short biographies of many well-known mathematicians

For ease of reading, a **second colour** is used to **highlight** each definition term. Within the definition text, a second colour is used to highlight terms that are defined elsewhere in the book. To ensure full understanding, students are directed to related terms that they may not understand.

Used as a ready reference, *Coming to Terms with Maths* will enhance both a student's understanding of mathematics, and their ability to perform independent revision.

Seamus McCabe

CONTENTS

KEY MA ... RMULAE
AND S ... AINED

Coming to Terms with MATHS

Seamus McCabe

EDCO

The Educational Company of Ireland
Ballymount Road
Walkinstown
Dublin 12

A trading unit of Smurfit Ireland Ltd

Editor: Michelle Gallen
Design and layout: Design Image
Cover design: Design Image
Cover images: Brian Fitzgerald and Design Image
Artwork: Brian Fitzgerald
Graphics: Seamus McCabe and Design Image

Printed in the Republic of Ireland by Future Print

0 1 2 3 4 5 6 7 8 9

Aa

abelian group A group in which the binary operation is commutative. It is also called a commutative group.

absolute error When an approximation (estimate) is made to a certain number (true value), absolute error is defined as follows:

absolute error = true value – estimate.

See also percentage error.

absolute value See modulus.

acceleration The rate of change of velocity with respect to time. Given a formula that expresses velocity as a function of time, then we can differentiate with respect to time to get a formula for acceleration. Acceleration is a vector quantity and the normal unit of measure is m/s^2.

acute angle An angle between 0° and 90° in measure. For example, 50° is an acute angle.
See also obtuse angle and reflex angle.

50°

adjacent One of the sides along the right angle in a right-angled triangle. The side adjacent to an angle forms that angle, together with the hypotenuse.

B
hypotenuse
adjacent to angle B
A
adjacent to angle A

Or

Adjacent sides in a quadrilateral are sides that meet at a point.

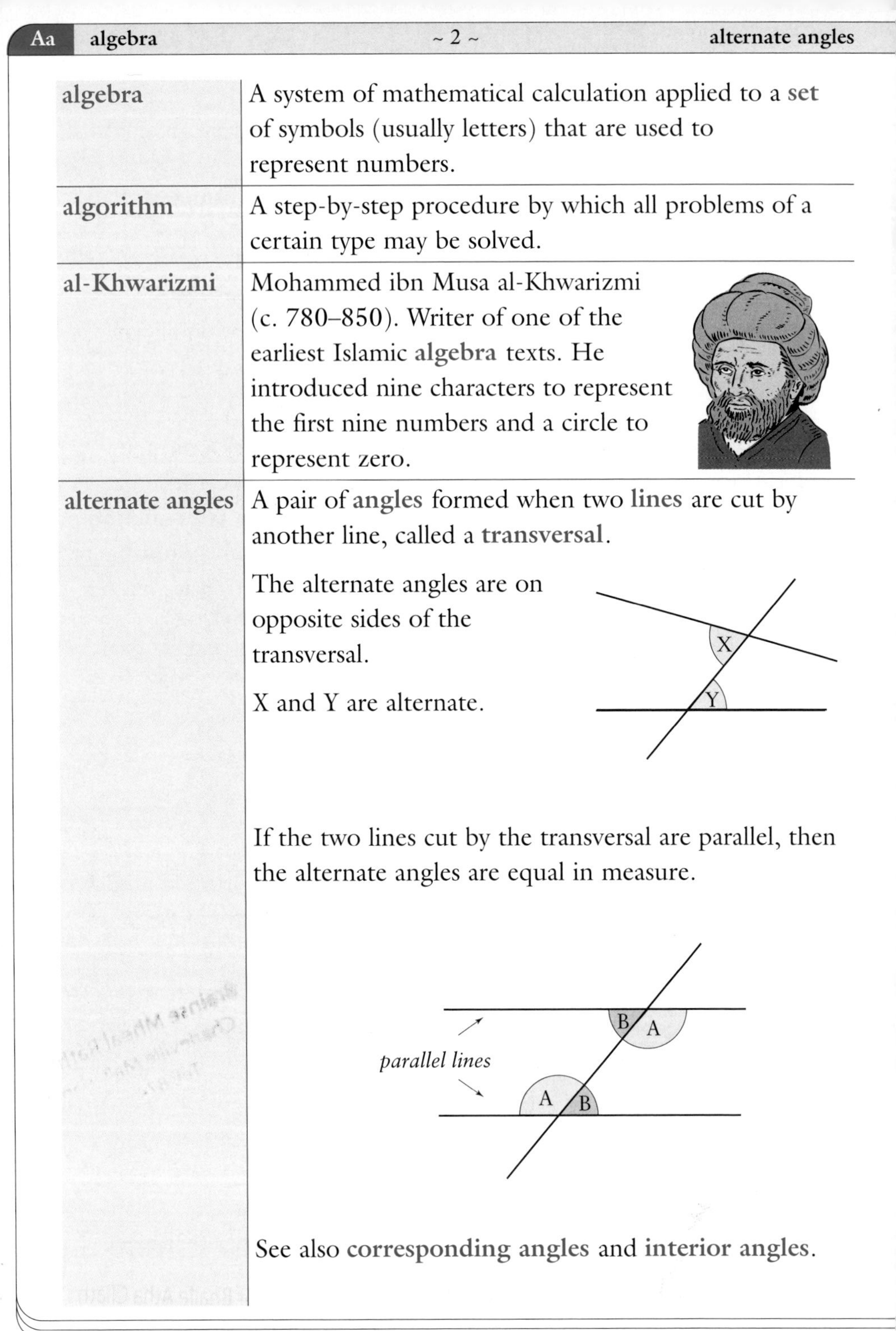

algebra	A system of mathematical calculation applied to a **set** of symbols (usually letters) that are used to represent numbers.
algorithm	A step-by-step procedure by which all problems of a certain type may be solved.
al-Khwarizmi	Mohammed ibn Musa al-Khwarizmi (c. 780–850). Writer of one of the earliest Islamic **algebra** texts. He introduced nine characters to represent the first nine numbers and a circle to represent zero.
alternate angles	A pair of **angles** formed when two **lines** are cut by another line, called a **transversal**. The alternate angles are on opposite sides of the transversal. X and Y are alternate. If the two lines cut by the transversal are parallel, then the alternate angles are equal in measure. See also **corresponding angles** and **interior angles**.

alternative hypothesis	In hypothesis testing, if the test statistic lies in the critical region at a given level of significance (see significance level), then we reject the null hypothesis and accept the alternative hypothesis. Consider the following problem: A die is thrown 600 times and the number 5 appears 120 times. At the 5% level of significance, determine whether or not the die is biased. The null hypothesis is that the die **is not** biased. The alternative hypothesis is that the die **is** biased. The null hypothesis is sometimes represented by H_0 and the alternative hypothesis by H_1.
altitude	See perpendicular height.
amount	The sum of money obtained when the interest is added to the principal if money is invested or borrowed. The formula $$A = P\left(1 + \frac{r}{100}\right)^n$$ is used to calculate the amount, A, where P is the sum of money invested or borrowed, r is the rate of interest (fixed) and n is the number of years. For example, €500 will amount to €540·80 (500×1.04^2) if it is invested for two years at 4% per year.
angle	An angle is formed when two lines intersect. The measure of the angle is the amount of rotation required to put one arm of the angle on top of the other arm. Angles are measured in degrees or radians.

angle between two lines formula

The formula $$\tan\theta = \pm\frac{m_1 - m_2}{1 + m_1 m_2}$$

is used in coordinate geometry to find the measure of the angle formed by two intersecting lines. To use the formula we need the slope of each line.

When $\tan\theta$ is a positive number, we have found the tan of the acute angle between the lines. When $\tan\theta$ is negative, we have found the tan of the obtuse angle.

If the lines are perpendicular, $\tan\theta$ is undefined and the denominator of the above fraction is zero. Therefore, when lines are perpendicular, $m_1 m_2 = -1$.

angle of depression

The angle formed by a line of vision and the horizontal, where the line of vision is in a downward direction.

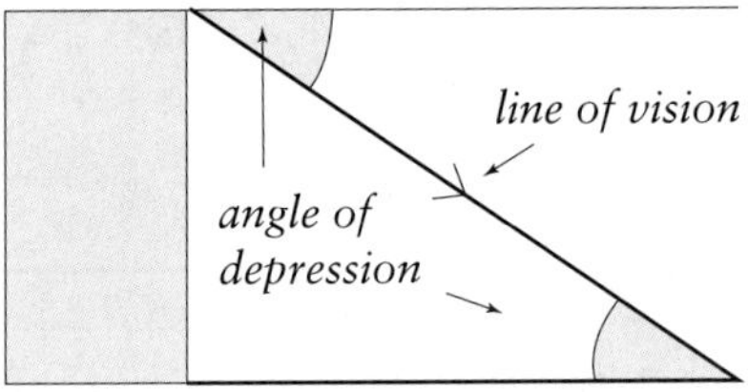

See also angle of elevation.

angle of elevation

The angle formed by a line of vision and the horizontal, where the line of vision is in an upward direction.

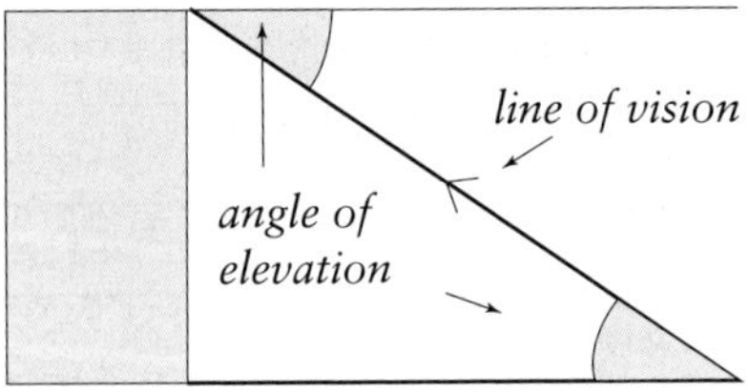

See also angle of depression.

appreciate	To increase in value. If an item costs €x and appreciates at the rate of r% per annum for n years, then its value after n years is given by the formula: $$\text{Value} = x\left(1 + \frac{r}{100}\right)^n$$ See also **depreciate** and **compound interest**.
arc	A continuous part of the **circumference** of a **circle**. A circle can contain three types of arc: a **semi-circle**, which is exactly half the circumference, a **minor arc**, which is less than half of the circumference and a **major arc**, which is more than half. 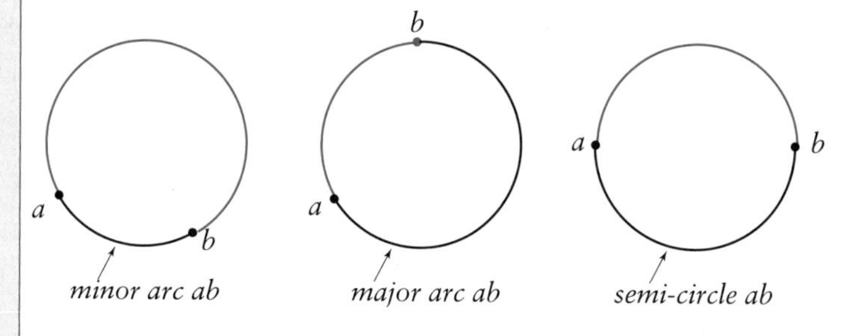
Archimedes	Archimedes (c. 287–212 BC). Greek mathematician and inventor regarded as the best of his day. He was born in Syracuse, Sicily, the son of an astronomer and studied in Egypt at Alexandria under the successors to **Euclid**.  He became famous for his design of many engines of war, for example, missile launchers. He developed an early form of **integration** for calculating areas and volumes. His most famous theory, entitled 'The Principle of Archimedes,' is that when an object is immersed in a liquid the volume of the object is equal to that of the displaced liquid.

arcsin	In **trigonometry**, the inverse (reverse) of the **sin function**, written $\sin^{-1}$. When using the calculator to evaluate arcsin, remember that $\sin^{-1} x$ is always between 0° and 90° when x is positive and between –90° and 0° when x is negative.
arctan	In **trigonometry**, the inverse (reverse) of the **tan function**, written $\tan^{-1}$. When using the calculator to evaluate arctan, remember that $\tan^{-1} x$ is always between 0° and 90° when x is positive and between –90° and 0° when x is negative.
area	The amount of space taken up by a **two-dimensional** object enclosed by a specific boundary. Area is measured in square units, for example, cm^2, m^2. See also **surface area** and **volume**.
Argand	Jean-Robert Argand (1768–1822 AD). Swiss-born bookkeeper and amateur mathematician credited with the development of a geometrical representation of complex numbers. See also **Argand diagram**.
Argand diagram	The name given to the diagram used to represent **complex numbers**. 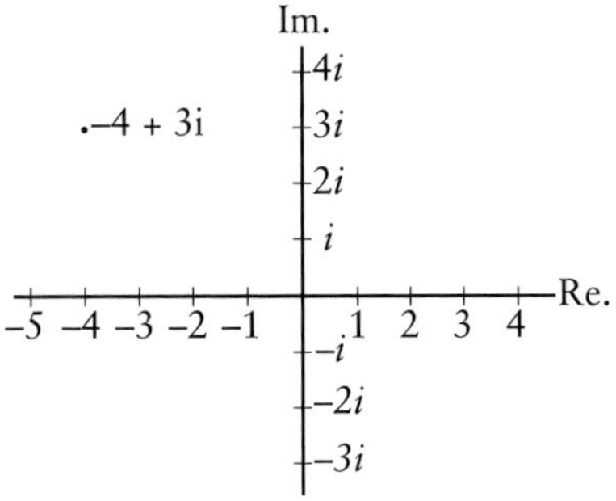 Real numbers are represented by dots on the horizontal axis. Imaginary numbers are represented by dots on the vertical axis. Numbers which have a real component and an imaginary component are represented by dots on a grid formed by the axes. Im. 4i 3i 2i i –i –2i –3i; Re. –5 –4 –3 –2 –1 1 2 3 4; •–4 + 3i The complex number $-4 + 3i$ is represented by the point $(-4, 3i)$. See also **modulus**, **argument** and **polar form**.

argument

The **angle** formed with the **positive sense** of the **real axis** when a **complex number**, z, is joined to **zero** on an **Argand diagram**. Often written as arg(z).

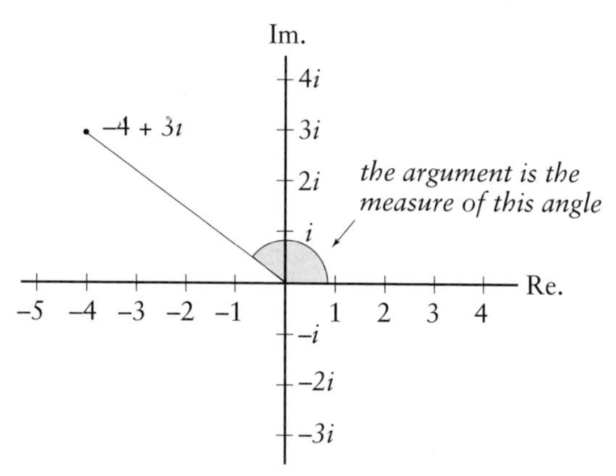

We usually take the argument to be a **negative angle** when the imaginary part of the complex number is negative.

Aristotle

Aristotle (384–322 BC). Born in Stagirus, in northern Greece. In 367 BC, at the age of seventeen, Aristotle became a student at Plato's Academy in Athens. He later taught there. In 335 BC, Aristotle founded a school in Athens, called the Lyceum.

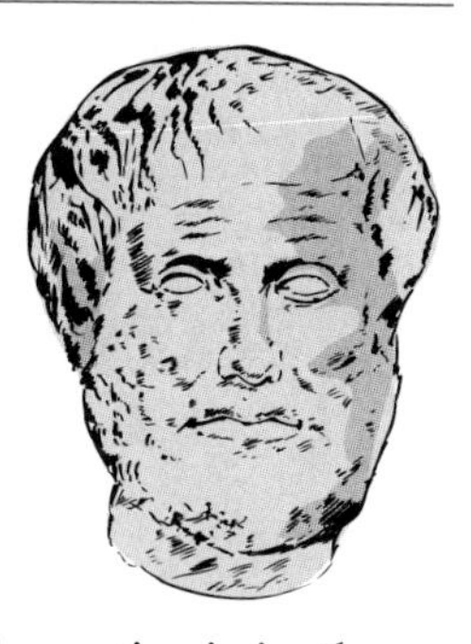

Aristotle's main contribution to mathematics is in the area of logic. He believed that logical arguments should be built from syllogisms. A syllogism is made up of some statements that are taken to be true leading to other statements that must then be true also.
For example:

Some animals are dogs.
All dogs have four legs.
Therefore, some animals have four legs.

arithmetic	The branch of mathematics concerned with numbers and their properties. The basic operations of arithmetic are addition, subtraction, multiplication and division.
arithmetic mean	The arithmetic mean of a **set** of numbers is found by adding the numbers and then dividing the answer by the amount of numbers. For example, the arithmetic mean of 1, 4, 6 and 9 is $\frac{1+4+6+9}{4} = 5.$ The arithmetic mean is also referred to as simply the mean or the average. See also **geometric mean**, **mode** and **median (statistics)**.
arithmetic sequence/series	A **sequence** or **series** is said to be arithmetic if the difference between consecutive terms is always the same (**constant**). This constant is referred to as the **common difference**. If the numbers are decreasing then the common difference is negative. For example: 3, 5, 7, 9, 11........... is an arithmetic sequence. 3 + 5 + 7 + 9 + 11........... is an arithmetic series. Notice that commas separate the numbers in the sequence, whereas the numbers in the series are added together.
arithmetico-geometric series	A type of **series** in which each term is made up of a **product**. The numbers in one half of the product are in **arithmetic sequence** whereas the numbers in the other half are in **geometric sequence**. For example, $1 + 3x + 5x^2 + 7x^3 + \ldots\ldots\ldots\ldots$

arms of an angle

The lines that form an angle.

arrangement

A collection of numbers or objects sorted in a particular order. For example, abc and bac are two different arrangements of the letters a, b and c. An arrangement is also called a permutation.

See also combination.

arrow diagram

A diagram used to represent a relation or a function.

If the relation is defined on a single set then the arrow diagram will involve one oval with arrows going from one element to another.

If the relation is defined between two sets, then the arrow diagram will involve two ovals with arrows going from one oval to the other.

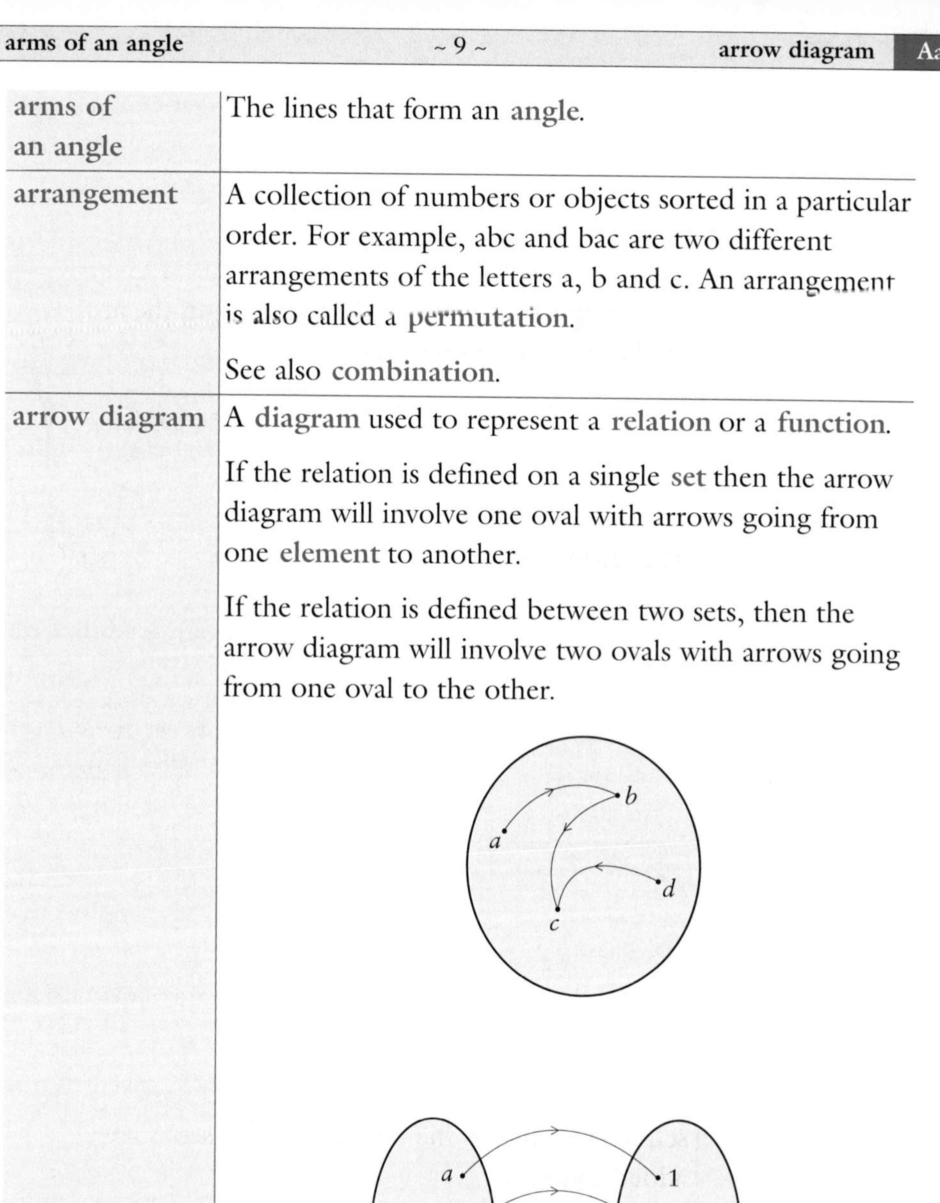

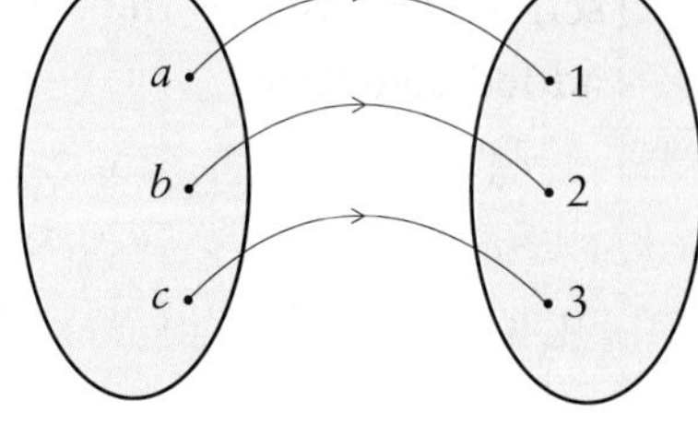

See also couple, domain and range.

associative	A binary operation, $*$, is said to be associative if $(a*b)*c = a*(b*c)$ for all a, b and c. Addition is associative, for example: $(2 + 5) + 7 = 2 + (5 + 7) = 14$ Subtraction is not associative since $(4 - 3) - 1 \neq 4 - (3 - 1)$ See also **commutative** and **distributive**.
assuming the converse	In a **theorem**, we are given x and asked to prove that y follows. The converse is that if we are given y then x follows. While the converse may be true in certain cases, it is incorrect to assume that it will always be true.
asymptote	A **line** that approximates a curve at its extremities. The diagram shows part of the graph of the function $y = \frac{1}{x}$. The x-axis and y-axis are asymptotes. 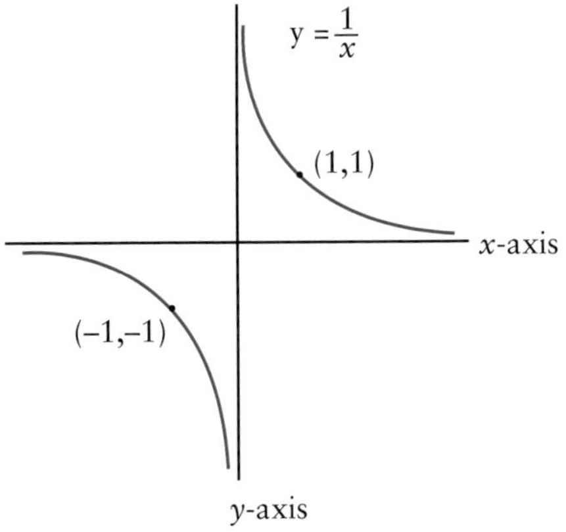 If the graph is continued, it will get closer and closer to the axes without ever touching them.

average	See **arithmetic mean**.
axial symmetry	Axial symmetry in a line L is a set of couples, (*a*, *b*), such that the **line** ***ab*** is **perpendicular** to L and the **midpoint** of the **line segment** [*ab*] is on L. Axial symmetry is also referred to as reflection in a line. 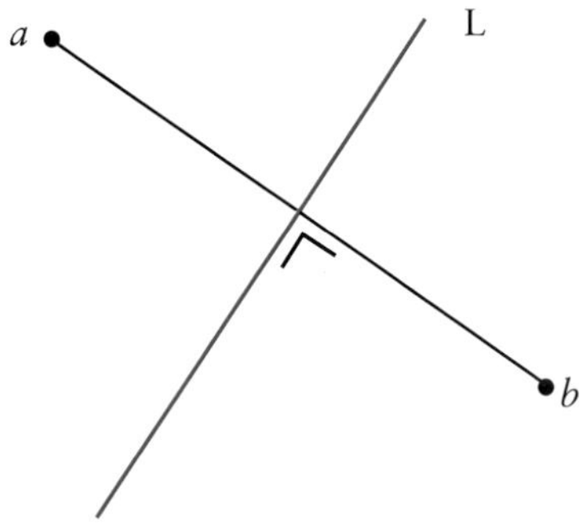 We say that *b* is the image of *a* under axial symmetry in the line L. An axial symmetry can be visualised by folding the page along the line L. The point *a* will fold onto its image point, *b*. See also **axis of symmetry**, **central symmetry** and **translation**.
axiom	A statement that can be taken to be true without proof. See also **theorem**.
axis	One of the lines used to locate points in **coordinate geometry**. In **two-dimensional** coordinate geometry, the horizontal line is called the *x*-**axis** and the vertical is the *y*-**axis**.

axis of symmetry	If an object can be folded so that one half fits exactly on top of the other half, then the line along which the object was folded is called an axis of symmetry. For example, a square has four axes of symmetry. A parallelogram has no axis of symmetry unless its sides are equal in length. 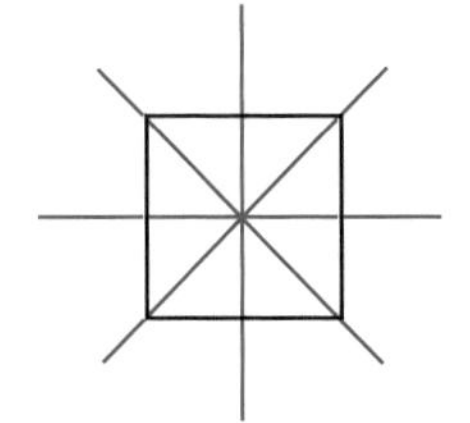

Bb

Babbage, Charles	Charles Babbage (1792–1871). English mathematician whose ideas led to the development of the computer. He designed the Difference Engine, followed by the Analytical Engine. Neither was completed. 
bar chart	A chart used in statistics to represent data. It consists of rectangles of equal width, separated by equal spaces. The bar chart can be horizontal or vertical. The height (or length) of each bar is proportional to the frequency. 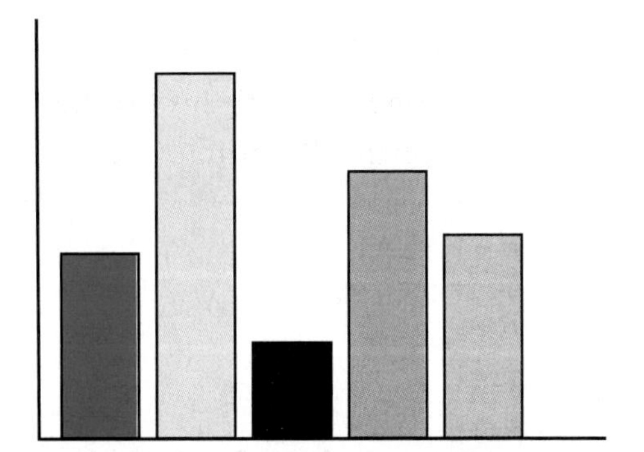 See also histogram, pie chart and trend graph.

base	In **geometry**, the base is one side of a **polygon**. In a **power**, for example, x^y, x is called the base. In a **log**, for example, $\log_a b$, a is called the base.
bearings	In **trigonometry**, bearings give us the location of one object relative to another. For example, the bearing of a from b is $N23°W$.
bell-shaped curve	See **normal curve**.
Berkeley, George	George Berkeley (1685–1753). Irish bishop and philosopher famous for his criticism of **Newton**'s work on **calculus**. His valid criticism led to clarifications from other mathematicians, most notably **Maclaurin**.
Bernoulli, Jacob	Jacob Bernoulli (1654–1705). Swiss mathematician after whom many results in probability and statistics are named. Together with his brother Johann Bernoulli (see **Bernoulli, Johann**), he was among the first mathematicians to attempt to understand and apply **Leibniz**'s theory of calculus. Later in life he was to develop a bitter rivalry with his brother. Jacob once wrote that Johann was his pupil and that his brother's only achievements were to repeat what his teacher had taught him.

Bernoulli, Johann	Johann Bernoulli (1667–1748). Swiss mathematician, brother of Jacob Bernoulli (see **Bernoulli, Jacob**). He was most famous for teaching **calculus** to Guillaume de **l'Hôpital** who proceeded to publish the first textbook on differential calculus based on those teachings. Many of Johann Bernoulli's discoveries were published in this book under l'Hôpital's name, but Johann remained silent on the matter since he had been well paid by l'Hôpital. The result in calculus known as **l'Hôpital's rule** is, in fact, the work of Johann Bernoulli. 
Bernoulli trial	When an **experiment** with only two possible outcomes (success or failure) is conducted repeatedly and the probabilities do not change between experiments, each experiment is called a Bernoulli trial.
biased	In **probability**, a die or coin is said to be biased if all the possible outcomes do not have an equal **probability** of success. See also **unbiased**.
bijection	A one-to-one correspondence. A bijection links each **element** of one **set** with one, and only one, element of another set.
binary operation	A means of combining two numbers or objects to produce a third. Addition, subtraction, multiplication, division, **union of sets** and **composition** of **transformations** are examples of binary operations.
binomial	In **algebra**, an **expression** made up of exactly two terms. For example, $5x - y$.

binomial theorem	In **algebra**, used to raise to a **power**, a bracket containing the sum or difference of two terms. The binomial theorem states that $$(x+y)^n = \binom{n}{0}x^n y^0 + \binom{n}{1}x^{n-1}y + \binom{n}{2}x^{n-2}y^2 + \binom{n}{3}x^{n-3}y^3 + \dots\dots\dots + \binom{n}{n}x^0 y^n$$ The general term of the **expansion** takes the form $$T_{r+1} = \binom{n}{r}x^{n-r}\,y^r$$ and is used to select a particular term without listing the entire expansion.
bisect	To cut exactly in half.
bisector of an angle	A **line** that cuts an **angle** in two equal parts. Every point on the bisector of an angle is **equidistant** from the **arms** of the angle. In a **rhombus**, the **diagonal** bisects the angles through which it passes.
bisector of an angle formula	$$\frac{a_1x + b_1y + c_1}{\sqrt{a_1^2 + b_1^2}} = \pm \frac{a_2x + b_2y + c_2}{\sqrt{a_2^2 + b_2^2}}$$ In **coordinate geometry**, the formula used to find the **equations** of the **lines** that **bisect** the **angle** formed by the lines $a_1x + b_1y + c_1 = 0$ and $a_2x + b_2y + c_2 = 0$.
Boole, George	George Boole (1815–64). English logician who taught at Queen's College in Cork from 1849 until his death. He is famous for Boolean **algebra**, which provides the logic behind the development of the modern-day computer. 

Cc

calculus	The branch of mathematics that uses the concept of **derivative** to analyse the way in which the value of a **function** varies. In the Leaving Certificate Higher Level course, calculus involves **differentiation** and **integration**. Calculus was developed by **Newton** and **Leibniz**, independently.
cancel	To remove **terms** from an **expression** to leave an identical, but simpler, expression. For example, $\frac{x^2 - y^2}{x - y} = \frac{(x - y)(x + y)}{(x - y)(1)} = x + y$
capacity	The **volume** inside an object, measured in cubic units.
cardinal number	The number of **elements** in a **set**. Represented by the symbol #. $\# \{1, 3, 5, 7\} = 4$
Cartesian	Relating to the work of **Descartes**.
Cartesian equation	An **equation** involving x and/or y. For example, $2x - 3y + 1 = 0$. See also **parametric equations**.
Cartesian form	A term used for one of the methods of representing a **complex number**. $x + iy$ is said to be in Cartesian form if x and y are real numbers and $i = \sqrt{-1}$. For example, $-\frac{1}{2} + \frac{\sqrt{3}}{2}i$ is in Cartesian form. See also **polar form**.
Cartesian plane	A **two-dimensional** space where the **points** are identified using **coordinates**.

Cayley table

A table used in groups to display the results obtained when all elements of the group are combined under the given binary operation. For example,

×	1	–1	i	$-i$
1	1	–1	i	$-i$
–1	–1	1	$-i$	i
i	i	$-i$	–1	1
$-i$	$-i$	i	1	–1

centimetre

Unit of measurement. Abbreviated to cm.

1 kilometre (km) = 1000 metres (m)
1m = 100 centimetres (cm)
1cm = 10 millimetres (mm)

central symmetry

Central symmetry in a point p is a set of couples, (a, b), such that the midpoint of the line segment $[ab]$ is p. Central symmetry is also referred to as reflection in a point.

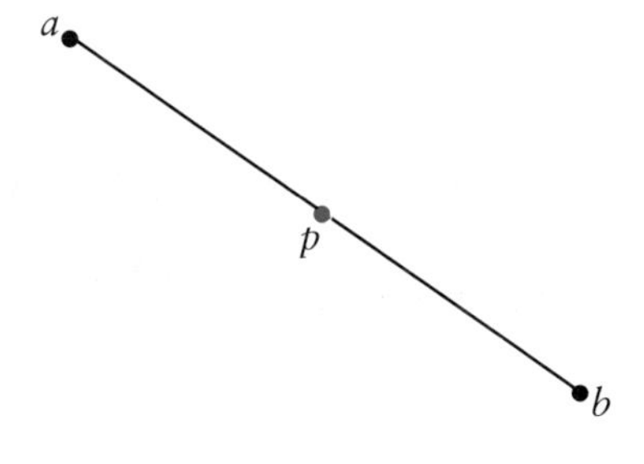

The point b is said to be the image of a under central symmetry in the point p.

See also centre of symmetry, axial symmetry and translation.

centre

A **circle** is a set of **points** that are **equidistant** from a fixed point.

This fixed point is called the centre.

circle

centre

See also **radius**.

centre of symmetry

If it is possible to **map** an object onto itself by a central **symmetry** in a point p, then p is called the centre of symmetry. For example, the centre of a **circle** is the centre of symmetry of a circle and the point of intersection of the **diagonals** is the centre of symmetry of a **parallelogram**.

centroid

In a **triangle**, the **point of intersection** of the medians. The centroid divides each **median (coordinate geometry)** in the **ratio** 2:1.

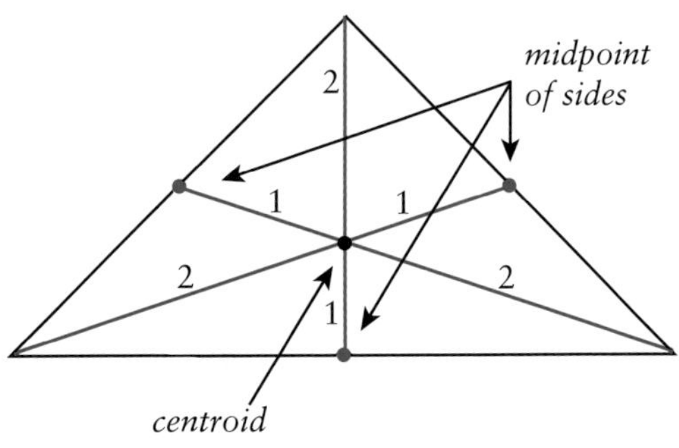

In **coordinate geometry**, the coordinates of the centroid may be obtained using the formula:

$$\left(\frac{x_1 + x_2 + x_3}{3}, \quad \frac{y_1 + y_2 + y_3}{3}\right)$$

In **vectors**, the centroid of the triangle abc is $\dfrac{\vec{a} + \vec{b} + \vec{c}}{3}$

chain rule	In **differentiation**, $\frac{dy}{dx} = \frac{dy}{du} \cdot \frac{du}{dx}$ The chain rule provides us with the following shortcut for differentiating a **function** that is raised to a **power**: $y = [f(x)]^n \Rightarrow \frac{dy}{dx} = n[f(x)]^{n-1}.f'(x)$ where $f'(x)$ is the **derivative** of $f(x)$. For example, $y = (x^3 - 1)^4 \Rightarrow \frac{dy}{dx} = 4\,(x^3 - 1)^3.3x^2$
chord	Any **line segment** joining two **points** on a curve. Of particular interest are chords on a **circle**. 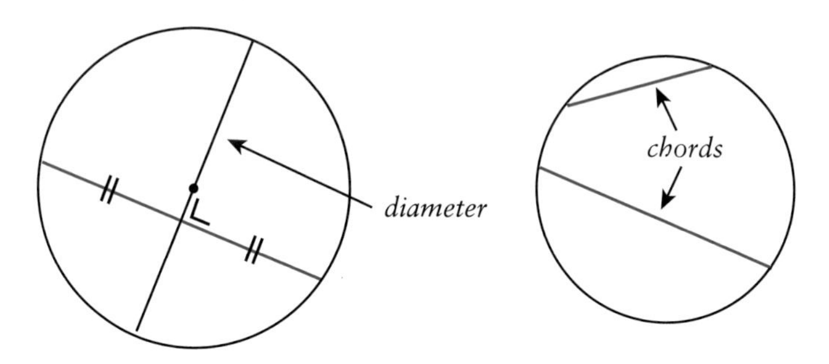 If a chord passes through the **centre**, it is called a **diameter**. If a diameter is drawn **perpendicular** to a chord, it will always **bisect** the chord. See also **secant** and **tangent**.
circle	A circle is a set of **points** that are **equidistant** from a fixed point called the **centre**. Note that the circle is made up of the points **on** the circumference, **not** the points **inside** the circumference. See also **radius**, **sector**, **segment**, **diameter**, **chord**, **secant** and **tangent**.

circumcentre	In a triangle, the point of intersection of the mediators (see diagram at circumcircle). The circumcentre is equidistant from the vertices of the triangle. It is the centre of the circumcircle. In a right-angled triangle, the circumcentre is the midpoint of the hypotenuse. See also centroid, incentre of a triangle and orthocentre.
circumcircle	A circle that passes through the vertices of a triangle. 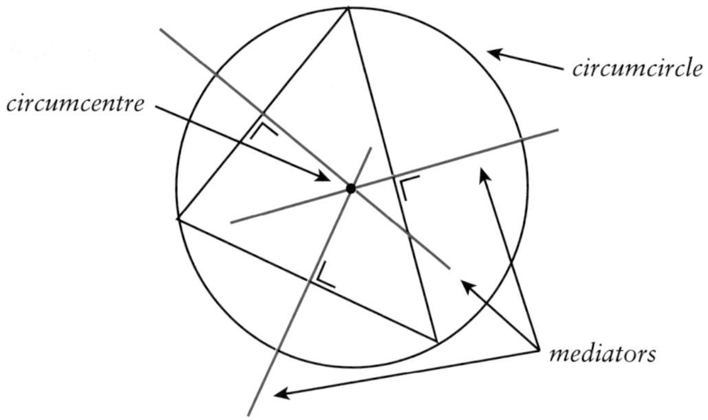
circumference	The curved line that forms a circle. The length of the circumference is given by the formula $2\pi r$, where r is the length of the radius of the circle.
closed	A set is said to be closed under a given binary operation if the result of combining any two elements of the set under the operation is always another element of the set. See also group.
cm	Abbreviation for centimetre.
coefficient	The number multiplying a variable in algebra. For example, in the expression $3x^2 - 5x + 1$, the coefficient of x^2 is 3 and the coefficient of x is -5.

collinear	In geometry, points are said to be collinear if they are on the same line. In coordinate geometry, we can prove that three points, a, b and c are collinear by showing that slope ab = slope bc.
column	A vertical array of numbers in a matrix.
column matrix	A matrix containing only one column.
combination	A selection where order does not matter. $\binom{n}{r} = \frac{n!}{r!(n-r)!}$ is the number of different combinations containing r objects, which can be made from n different objects. See also permutation.
common chord	A line segment that joins the points of intersection of two circles. *common chord*
common denominator	The common denominator of two or more fractions is a number into which all the denominators will divide evenly. In order to add fractions, there must be a common denominator. For example, in $\frac{3}{4} + \frac{1}{5} = \frac{15}{20} + \frac{4}{20}$ $= \frac{19}{20}$ the common denominator is 20. See also lowest common multiple.

common difference	The number obtained by subtracting any **term** from the next term in an **arithmetic sequence/series**. For example, in the arithmetic sequence 1, 5, 9, 13....... the common difference is 4. When the terms are decreasing, the common difference is negative.
common factor	A number that divides evenly into two or more numbers is a common **factor** of those numbers. For example, 4 is a common factor of 8 and 20. See also **highest common factor**, **multiple** and **lowest common multiple**.
common log	A **logarithm** whose **base** is 10. See also **natural log (logarithm)**.
common multiple	A number into which two or more numbers divide evenly is a common **multiple** of those numbers. For example, common multiples of 3 and 4 are: 12, 24, 36 .... See also **lowest common multiple**, **factor** and **highest common factor**.
common ratio	The number obtained by dividing any **term** by the previous term in a **geometric sequence/series**. For example, in the geometric sequence $1, \frac{1}{2}, \frac{1}{4}, \frac{1}{8},$ .... the common ratio is $\frac{1}{2}$.
commutative	A **binary operation** * is said to be commutative if $a*b = b*a$ for all a and b. For example, addition is commutative since $a + b$ is always equal to $b + a$. Subtraction is not commutative since $2 - 1 \neq 1 - 2$. See also **associative** and **distributive**.
commutative group	See **abelian group**.
complement	The complement of a set A is the set of **elements** that are in the **universal set** but are not in A. It is written A'.

complementary angles	Two **angles** that add up to 90°. For example, 20° and 70° are complementary angles. More generally, A and $90° - A$ are complementary angles. A **right-angled triangle** contains a **right angle** and two complementary angles. Note that $\cos(90° - A) = \sin A$ and $\sin(90° - A) = \cos A$ See also **supplementary angles**.
complex conjugate	In **complex numbers**, the conjugate of $a + bi$ is $a - bi$. If z is a complex number, then its conjugate is written $\bar{z}$.
complex fraction	See **compound fraction**.
complex numbers	Numbers that can be written in the form $a + bi$, where a and b are **real numbers** and $i = \sqrt{-1}$. See also **Argand diagram** and **polar form**.
composite number	A **natural number** that is bigger than 1 and has more than two **factors**. For example, 4, 6, 8, 9.........
composition	The composition of two **transformations** is a single transformation that produces the same result as one transformation followed by another. For example, if **axial symmetry** in the x-**axis** is followed by axial symmetry in the y-**axis**, the result is the same as that obtained by a **central symmetry** in the origin. For example: $S_y \circ S_x = S_O$
compound angle	An **angle** involving a **sum** or a difference, for example, $A + B$ or $A - B$.
compound angle formulae	In **trigonometry**, the formulae that convert **sin**, **cos** or **tan** of a compound angle $A + B$ into an expression involving the sin and cos of A and B. The formulae for the compound angle $A + B$ is: $\cos(A + B) = \cos A \cos B - \sin A \sin B$ $\sin(A + B) = \sin A \cos B + \cos A \sin B$ $\tan(A + B) = \dfrac{\tan A + \tan B}{1 - \tan A \tan B}$ can be found on **page 9** of the mathematical tables.

compound fraction	A **fraction** that contains other fractions. For example, $\dfrac{\frac{2}{3}+1}{\frac{1}{2}}$. Also called a **complex fraction**. The compound fraction can be simplified by multiplying the top and bottom by the common denominator of all the fractions involved. $\dfrac{\frac{2}{3}+1}{\frac{1}{2}} \begin{matrix}\times 6\\ \times 6\end{matrix} = \dfrac{4+6}{3} = \dfrac{10}{3}$
compound inequality	A statement containing two **inequality** symbols. For example, $x - 3 \leq 2x + 1 \leq 5x - 2$
compound interest	When a sum of money is invested in a bank, it earns **interest**. The interest earned is a certain **percentage** of the sum invested. In the second year, this interest is added to the original sum and the interest for the second year is a percentage of this new figure and so on. The total interest is called compound interest. See also **appreciate** and **depreciate**.
compound interest formula	$A = P\left(1 + \dfrac{r}{100}\right)^n$ where A is the total value of the investment, including compound interest, P is the sum of money invested, r is the rate of interest and n is the number of years for which the sum of money is invested.
concentric	Geometric figures, usually circles, that have the same centre.
conclusion	A final statement, arrived at by means of a **proof**.

concurrent lines

Lines that have the same point of intersection.
In coordinate geometry, the formula

$$\mu L_1 + \lambda L_2 = 0$$

gives the equation of every line that is concurrent with the lines

$$L_1 = 0 \text{ and } L_2 = 0.$$

concyclic

A number of points are concyclic if it is possible to draw a single circle that passes through them all.

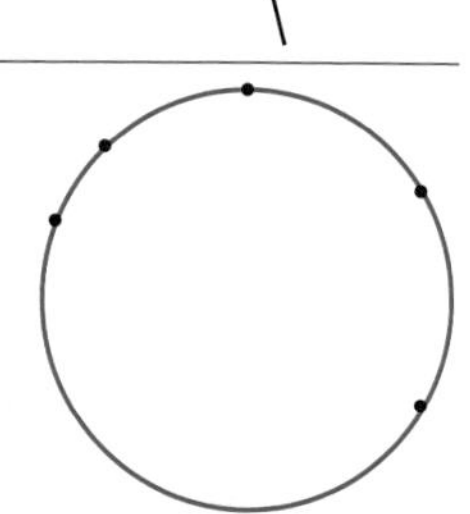

conditional probability

When the probability that a particular event will occur is influenced by the fact that another event has already occurred.

In this case, $P(E|F) = \dfrac{P(E \cap F)}{P(F)}$, where $P(E|F)$ is the probability that E will occur given that F has occurred.

For example, a class consists of twenty girls and ten boys. Seven girls and two boys wear glasses.
One student is chosen at random.

(a) What is the probability that the student wears glasses?

(b) What is the probability that the student wears glasses, *given that the student is a girl.*

Answer: (a) $\dfrac{9}{30}$ (b) $\dfrac{7}{20}$

Part (b) involves conditional probability because the answer is influenced by the fact that the chosen student is a girl.

See also dependent events and independent events.

cone	A **three-dimensional** figure with a circular or elliptical base, tapering to a fixed point called the **vertex**. When we use this word we usually mean a right-circular cone, i.e. one with a circular base in which a line from the vertex to the centre of the circle is **perpendicular** to the **plane** of the **circle**. The volume of the cone is $\frac{1}{3}\pi r^2 h$ The curved surface area is πrl The total surface area is $\pi rl + \pi r^2$ Also, $r^2 + h^2 = l^2$ 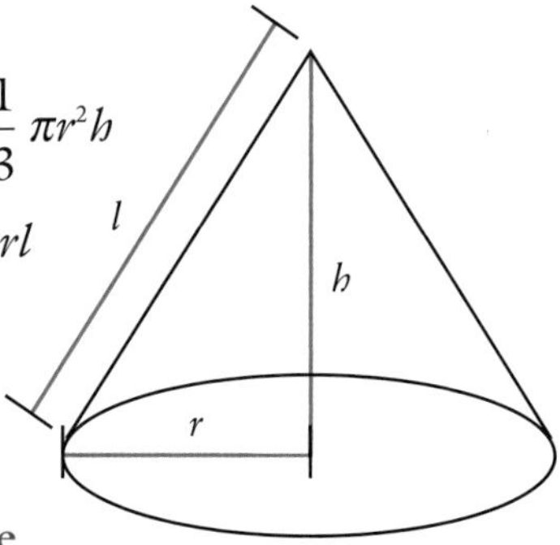 See also **cylinder** and **sphere**.
confidence interval for the mean	A **range** of values in **sampling theory** within which we expect to find the **mean**. The 95% confidence interval for the mean, μ, is $\bar{x}_1 - 1.96\sigma_{\bar{x}} < \mu < \bar{x}_1 + 1.96\sigma_{\bar{x}}$ where $\bar{x}_1$ is the mean of a sample and $\sigma_{\bar{x}}$ is the **standard error**.
confidence level	A confidence level measures the dependability of a result. The higher the confidence level, the more likely it is that the result is accurate.
congruent triangles	**Triangles** that are identical in every respect. To prove that two triangles are congruent, we must identify three parts (sides or **angles**) that we know to be the same in both triangles. The three parts must fit into one of the following categories: SSS – three sides of one triangle are equal to the three sides of the other. SAS – two sides and the angle where they meet. ASA – two angles and the side which they share. RHS – the triangles are right-angled, the hypotenuse is the same in each and they have another equal side. See also **equiangular triangles**.

conjugate surd	The conjugate of a compound surd $a + \sqrt{b}$ is $a - \sqrt{b}$. See also **complex conjugate**.
constant	A numerical part of an **expression** in **algebra**. For example, in the expression $x^3 - 3x^2 + x + 5$, the constant is 5. See also **coefficient**.
constant of integration	A constant, usually denoted by C, that forms part of the solution in an **indefinite integral**. For example: $\int x^2\, dx = \frac{1}{3}x^3 + C$ See also **definite integral**.
constraint	A condition that applies to a particular problem. For example, in the problem 'How many four-digit numbers can be formed using the digits 1, 2, 3, 4, 5 and 6, if no digit may be repeated,' the constraint is 'no digit may be repeated.'
construct	To draw exactly with the use of mathematical instruments.
construction	A figure that has been drawn with the use of mathematical instruments. Or An addition made to a diagram in the course of proving a **theorem**.
continuity correction	In **probability**, an adjustment made when we use the **normal distribution** to approximate the **binomial distribution**. It involves either adding 0.5 to, or subtracting 0.5 from, the **mean** to compensate for the fact that we are using a curve that applies to a continuous distribution to answer a question involving a discrete distribution.
continuous random variable	In **probability**, the distribution of a random variable (x) is continuous when the variable can take any value in a certain range. See also **discrete random variable**.

convergent	An **infinite series** is convergent if there is a **limit** to how big the overall sum becomes. Mathematically, we say that $\lim_{n \to \infty} S_n$ exists. The number to which a series converges is called the sum to infinity of the series. See also **divergent** and **ratio test**.
converse	In a **theorem** we are given A to prove B. The converse is when we are given B to prove A.
coordinate geometry	An alternative approach to the **geometry** of **Euclid** in which every **point** in the **plane** is given **coordinates** and the properties of **plane objects** can then be studied using **algebra**.
coordinates	In **coordinate geometry**, coordinates are used to specify the location of a **point**. The diagram shows the location of the point with coordinates (2, –3). 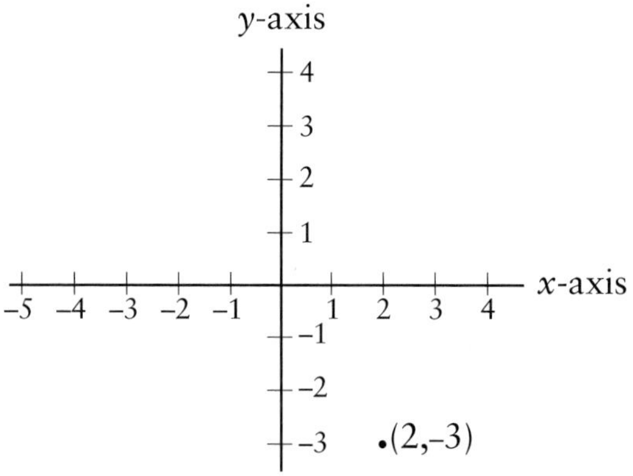 *y*-axis *x*-axis •(2,–3)
corollary	A result that follows directly from the proof of another result. For example, a **theorem** proves that the measure of an angle at the **centre** of a **circle**, standing on an **arc**, is twice the measure of any angle at the circle, standing on the same arc. A corollary is that all angles at the circle, standing on the same arc, are equal in measure.

corresponding angles	When two **lines** are cut by a **transversal**, corresponding angles are formed. They are on the same side of the transversal and either above or below each **line**. If the two lines are **parallel**, then corresponding angles are equal. Angles represented by the same number are corresponding. See also **alternate angles** and **interior angles**. Or Angles that are opposite the same side in **congruent triangles**. They are equal in measure.
corresponding sides	Sides that are opposite the same angle in **congruent triangles**. They are equal in length.
cos	Short for cosine. It is the ratio adjacent:hypotenuse in a **right-angled triangle**. In the diagram, $\cos A = \frac{x}{z}$ See also **sin**, **tan**, **cosec**, **sec** and **cot**.
cosec	Short for cosecant. It is the ratio of hypotenuse:opposite in a **right-angled triangle**. Note: $\text{cosec}A = \frac{1}{\sin A}$ In the diagram, $\text{cosec}A = \frac{z}{y}$ See also **sin**, **cos**, **tan**, **sec** and **cot**.

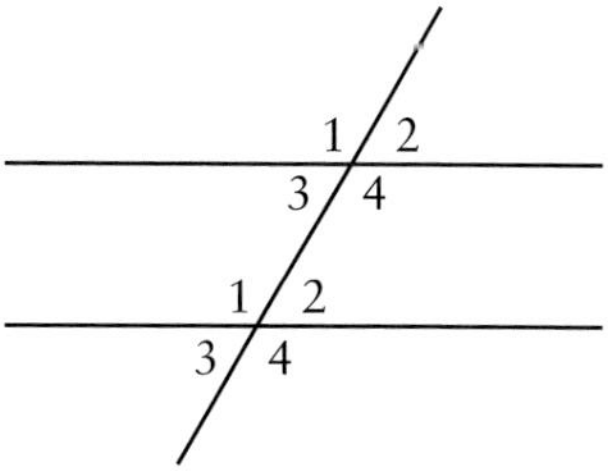

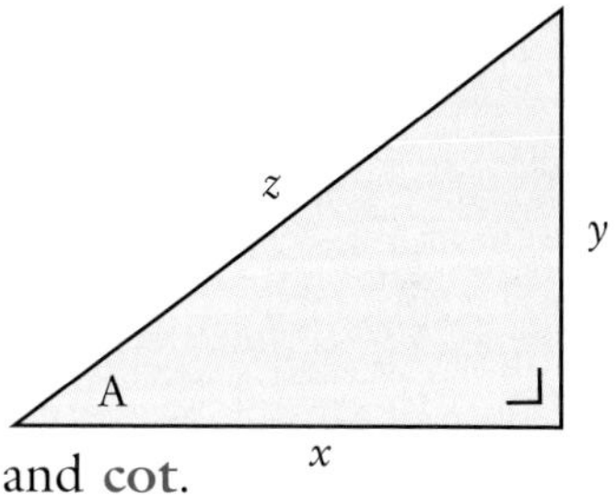

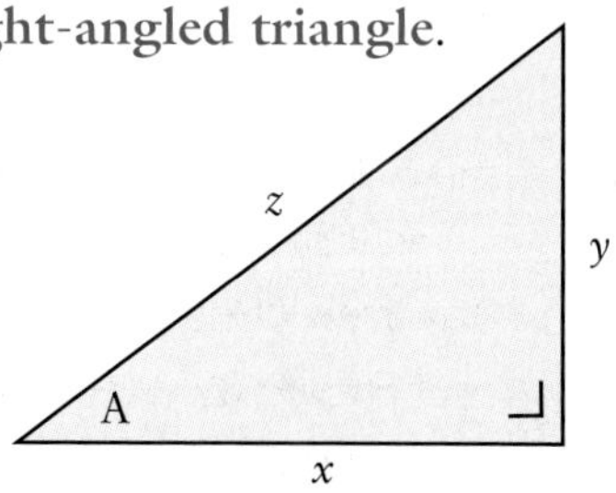

cosine rule	In a **triangle**, the cosine rule $a^2 = b^2 + c^2 - 2bc \cos A$ is used to find the length of the side a when we know the lengths of the other two sides and the measure of the **angle** between them. The cosine rule can be rearranged to read $\cos A = \frac{b^2 + c^2 - a^2}{2bc}$ and used to find the measure of the angle A when we know the lengths of the three sides. See also **sine rule**.
cost price	The price of an article before profit is added. Cost Price + Profit = Selling Price.
cot	Short for cotangent. It is the ratio of adjacent:opposite in a **right-angled triangle**. Note: $\cot A = \frac{1}{\tan A} = \frac{\cos A}{\sin A}$ z y A x In the diagram, $\cot A = \frac{x}{y}$ See also **sin**, **cos**, **tan**, **sec** and **cosec**.
couple	An ordered pair. For example, (a, b). Note: $(a, b) \neq (b, a)$ See also **relation** and **function**.
critical region	In **probability**, the set of values for which the **null hypothesis** is rejected. See also **hypothesis testing** and **significance level**.
critical standard score	The **standard score** that defines the **critical region** in **hypothesis testing**. For a one-tailed test it is 1.645. For a two-tailed test it is 1.96.

cube	Three-dimensional box with all sides equal in length. If x is the length of each side, then Volume = x^3 Surface Area = $6x^2$ 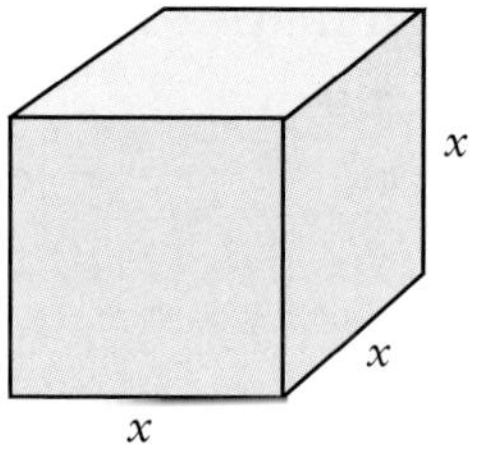
cube root	The cube root of n is a number that gives n when it is raised to the power of three (cubed). For example, the cube root of 8 is 2, since $2^3 = 8$. Mathematically, we write $\sqrt[3]{8} = 2$. $\sqrt[3]{x}$ can also be written $x^{1/3}$. See also square root.
cubic equation	An equation in one variable where the highest power of the variable is 3. For example, $x^3 - 2x^2 + x - 4 = 0$ is a cubic equation. See also quadratic equation.
cubic expression	An expression in one variable where the highest power of the variable is 3. For example, $2x^3 - x^2 + 7x - 1$ is a cubic expression. See also quadratic expression and coefficient.
cumulative frequency curve	A graph drawn from a cumulative frequency table with frequency on the vertical axis. Also called an ogive. The cumulative frequency table on page 32 is represented as follows: 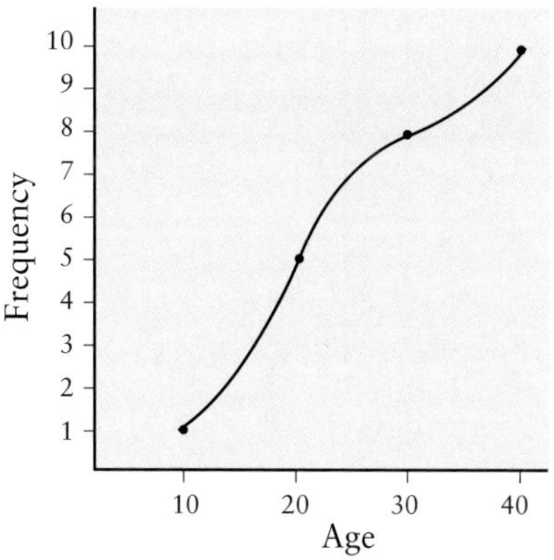

cumulative frequency table

In statistics, a table that is created from a grouped frequency distribution table where each category has an upper limit only. For example, a cumulative frequency table might show the number of people below certain ages, as follows:

Age	<10	<20	<30	<40
Frequency	1	5	8	10

cyclic group

A group, every element of which is a power of a single element of the group called the generator.

cyclic quadrilateral

A quadrilateral whose vertices are on the circumference of a circle.

The opposite angles in a cyclic quadrilateral add up to 180°, i.e. $A + B = 180°$.

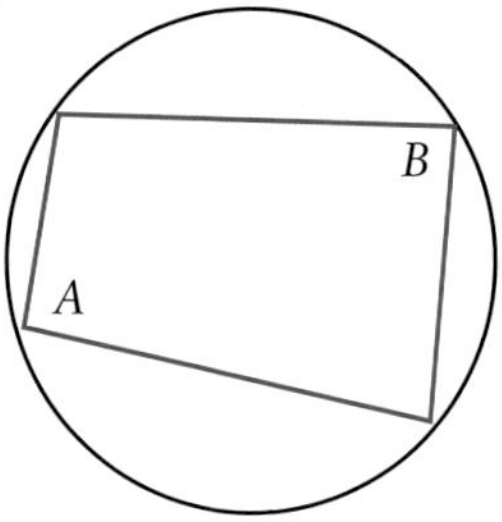

cylinder

A tubular three-dimensional object with a circular base. When we use the word cylinder, we usually mean a right cylinder, i.e. one in which the curved surface is perpendicular to the base.

The volume of a cylinder is $\pi r^2 h$.
The curved surface area is $2\pi rh$.
The total surface area is $2\pi rh + 2\pi r^2$.

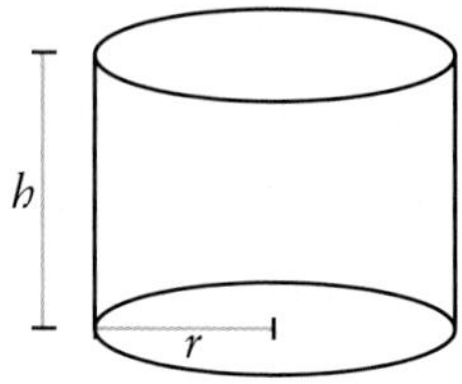

See also sphere, hemisphere and cone.

Dd

data	The raw facts and figures collected in statistics.
de Moivre	See de Moivre, under M.
de Moivre's theorem	See de Moivre's theorem, under M.
decimal	A fraction where the bottom number is a power of 10. It is written using a dot called a decimal point. For example, $\frac{6}{10} = 0.6$, $\frac{17}{100} = 0.17$, etc. Whole numbers are written to the left of the decimal point. For example, $1\frac{3}{10} = 1.3$.
decimal places	The number of figures (including zeros) to the right of the decimal point in a number is the number of decimal places in that number. For example, the number 23.274 has three decimal places and is between 23.27 and 23.28. When asked to round a number correct to two decimal places, choose the closest number that contains only two decimal places. For example, 23.274 is between 23.27 and 23.28 and becomes 23.27. 23.275 is exactly half-way between 23.27 and 23.28. When that happens, always choose the larger number, in this case, 23.28. See also significant figures.
decimal point	The dot that separates whole numbers from fractions when a number is written using the decimal system. In the number 17.23, the whole number is 17 and 23 represents the fraction $\frac{23}{100}$

decreasing function	A function $f(x)$ is said to be decreasing if $f(a) < f(b)$ for all $a > b$. When a function is decreasing, its derivative is negative. See also **increasing function**.
deduction	A form of **proof** used, for example, in **geometry**. Deduction is the procedure whereby we start out with certain **premises** and argue towards a **conclusion** that must be true so long as the initial premises are true. For example, if $a = b$ and $b = c$ then $a = c$. See also **induction**.
definite integral	An integral in the form $\int_a^b f(x)dx$. $\int_a^b f(x)dx = F(b) - F(a)$, where $F(x) = \int f(x)dx$ See also **indefinite integral**.
definition	A precise statement of meaning.
degree	A unit of measure for **angles**, the symbol for which is °. Rotation through a full circle is divided into 360 equal steps, each of which is one degree. Therefore, a circle contains 360°. See also **radian**. Or The degree of an **equation** is determined by the highest **power** of the **variable** involved. For example, $x^3 - x = 0$ is an equation of degree 3.
denominator	The bottom number in a **fraction**. See also **numerator** and **rational numbers**.
dependent events	In **probability**, two events, E and F, are dependent if the probability of one occurring is influenced by whether or not the other has already occurred. If E and F are dependent, $P(E \cap F) = P(E).P(F\|E)$. See also **independent events**.

depreciate	To lose value. If an item costs €x and depreciates at the rate of r% per annum for n years, then its value after n years is given by the formula $$\text{Value} = x\left(1 - \frac{r}{100}\right)^n$$ See also **appreciate** and **compound interest**.
depreciation	When an item loses value, the depreciation is the difference between its initial value and its final value.
derivative	The **limit** of the **slope** of a **chord**, linking two **points** on a curve, as the distance between the points tends to **zero**. Defined mathematically, the derivative $$f'(x) = \lim_{h \to 0} \frac{f(x+h) - f(x)}{h}$$ The derivative is the result obtained when we differentiate and is often denoted $\frac{dy}{dx}$.
Descartes	Rene Descartes (1596–1650). French mathematician and philosopher credited with founding **coordinate geometry**. See also **Cartesian equation**, **Cartesian form** and **Cartesian plane**.
determinant	Used to find the inverse of a **matrix**. The inverse of $\begin{pmatrix} a & b \\ c & d \end{pmatrix}$ is $\frac{1}{ad - bc}\begin{pmatrix} d & -b \\ -c & a \end{pmatrix}$ $ad - bc$ is called the determinant since its value determines whether or not the matrix has an **inverse**. If $ad - bc = 0$, then the matrix does not have an inverse and is called **singular**.

deviation	In statistics, the difference between an individual number and the mean. It is obtained by subtracting the mean from the number. See also standard deviation.
diagonal	A line segment joining two vertices of a polygon, which is not one of the sides. In a parallelogram, the diagonals bisect each other.
diagonal matrix	A matrix, all of whose entries are zero, except for those on the diagonal from the top left to bottom right. For example, $\begin{pmatrix} 2 & 0 \\ 0 & -3 \end{pmatrix}$
diagram	A picture used to represent a mathematical situation. For example, a Venn diagram.
diameter	A chord of a circle that passes through the centre. The length of the diameter is twice the length of the radius.
difference equation	A definition of a sequence that describes how to find the value of any term by using previous terms. It is sometimes called a recurrence relation. For example, $u_{n+2} = u_{n+1} + u_n$ A difference equation is a recursive definition of a sequence. To solve a difference equation is to replace the recursive definition with a more direct definition that expresses u_n as a function of n. This allows us to identify any term of the sequence without reference to previous terms.
difference of two cubes	An expression in the form $x^3 - y^3$ whose factors are $(x - y)(x^2 + xy + y^2)$. See also sum of two cubes and difference of two squares.

difference of two squares	An expression in the form $x^2 - y^2$ whose factors are $(x - y)(x + y)$. See also difference of two cubes and sum of two cubes.
differentiation	The process of evaluating the derivative.
digit	Any one of the figures 0, 1, 2, 3, 4, 5, 6, 7, 8 or 9.
direct proportion	Two numbers a and b are in direct proportion to two others c and d if $\frac{a}{b} = \frac{c}{d}$ See also inverse proportion.
discount	A reduction in price that is expressed as a percentage of the selling price. For example, if an article is reduced from €80 to €64 in a sale, then the discount is 20%. See also profit and loss.
discrete random variable	In probability, the distribution of a random variable (x) is discrete when the variable can only take certain values in a range. For example, a die is rolled 100 times and x represents the number of sixes that appear. x is a discrete random variable since it can only be a whole number between 0 and 100. See also continuous random variable.
discriminant	The name given to the expression $b^2 - 4ac$, which appears in the formula $x = \frac{-b \pm \sqrt{b^2 - 4ac}}{2a}$ which is used to solve quadratic equations in the form $ax^2 + bx + c = 0$. The value of the discriminant tells us the nature of the roots: $b^2 - 4ac \geq 0 \Rightarrow$ the roots are real $b^2 - 4ac < 0 \Rightarrow$ the roots are imaginary $b^2 - 4ac = 0 \Rightarrow$ the roots are identical

distance formula	$\sqrt{(x_2 - x_1)^2 + (y_2 - y_1)^2}$. Used in **coordinate geometry** to find the distance between the points (x_1, y_1) and (x_2, y_2). See also **perpendicular distance formula**.
distinct	Not the same number.
distributive	One binary operation, *, is said to be distributive over another, ⊗, if a * (b ⊗ c) = (a * b) ⊗ (a * c), for all a, b and c. For example, multiplication is distributive over addition since $x(y + z) = xy + xz$. See also **associative** and **commutative**.
divergent	An **infinite series** is divergent if there is no **limit** to how big the overall sum becomes. For example, the infinite series $1 + \frac{1}{2} + \frac{1}{3} + \frac{1}{4} + \ldots\ldots..$ is divergent, so by adding a sufficient number of terms we can make the overall sum as big as we wish. See also **convergent** and **ratio test**.
division in a ratio formula	The **formula** for dividing a **line segment** $[ab]$ internally in the ratio $m{:}n$ is: $p\left(\frac{mx_2 + nx_1}{m + n}, \frac{my_2 + ny_1}{m + n}\right)$ 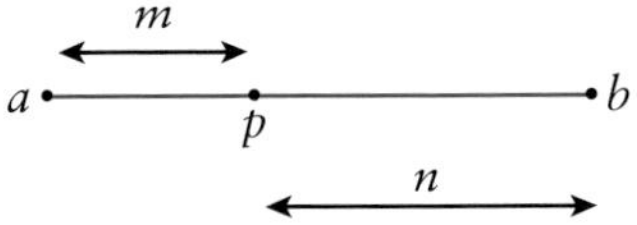 where $a\,(x_1, y_1)$ and $b(x_2, y_2)$. The formula for dividing a line segment $[ab]$ externally in the ratio $m{:}n$ is: $p\left(\frac{mx_2 - nx_1}{m - n}, \frac{my_2 - ny_1}{m - n}\right)$ 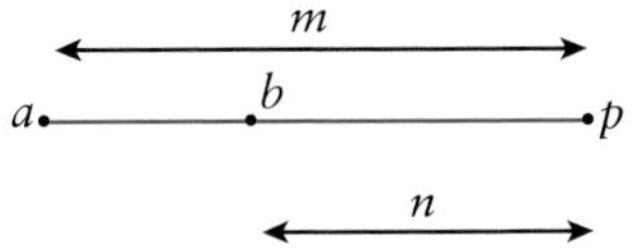 where $a\,(x_1, y_1)$ and $b(x_2, y_2)$. See also **midpoint formula**.
divisor	See **factor**.

domain	The set of first components in a relation or function. For example, in the relation $\{(1, 1), (2, 4), (3, 9), (4, 16)\}$, the domain is {1, 2, 3, 4). See also range.
dot product	In vectors, $\vec{a}.\vec{b} = \lvert\vec{a}\rvert.\lvert\vec{b}\rvert \cos\theta$, where $\lvert\vec{a}\rvert$ is the length of the vector $\vec{a}$ and θ is the angle between vectors $\vec{a}$ and $\vec{b}$. Also called scalar product as it produces a scalar. Of particular importance is the fact that the dot product of perpendicular vectors is zero.
double-angle formula	In trigonometry, a formula that expresses the value of a function of an angle in terms of functions of half the angle. For example: $\sin 2A = 2 \sin A \cos A, \qquad \cos 2A = \frac{1 - \tan^2 A}{1 + \tan^2 A}$ A double-angle formula can be applied to any angle. For example: $\cos\theta = 1 - 2\sin^2 \frac{\theta}{2}$. See also half-angle formula.

Ee

e	The transcendental number $e = \lim_{n\to\infty}\left(1 + \frac{1}{n}\right)^n$. It is approximately 2.718. See also exponential function.
Einstein	Albert Einstein (1879–1955). German physicist responsible for the Special Theory of Relativity. He was awarded the Nobel Prize in Physics in 1921.

element	One of the objects that make up a **set**. The symbol ∈ means 'is an element of' and the symbol ∉ means 'is not an element of.' When listing the elements of a set, use chain brackets and separate the elements using commas, for example, {2, 3, 5, 7}. An element of a set is also referred to as a **member** of the set.
empty set	A **set** that contains no **elements**. It is also called the null set. The symbol { } or Ø represents the empty set. On a **Venn diagram**, if a section is empty it is usually shaded.
enlargement	A **transformation** whereby an object is either increased or reduced in size by drawing lines from a fixed point (centre of enlargement) through each **vertex** of the object. 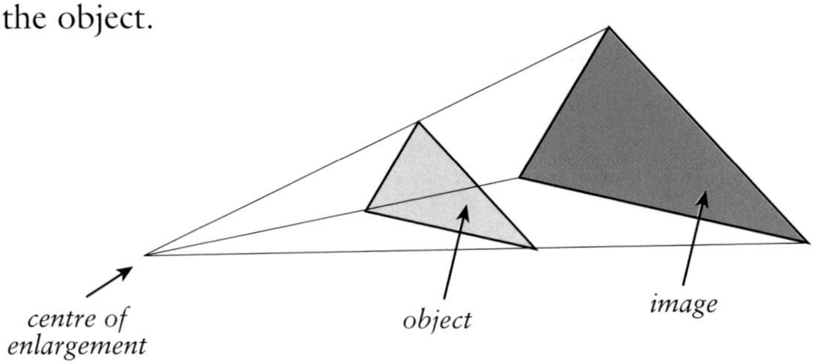 If the image is closer to the centre of enlargement than the object, then the object will be reduced in size. 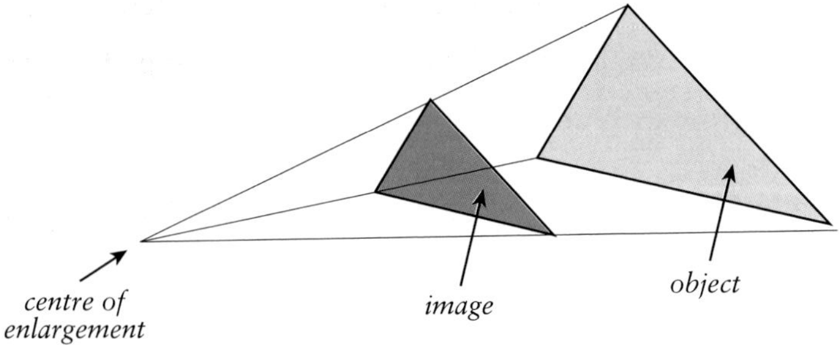 See also **scale factor**.

equal sets	Sets that contain exactly the same elements, regardless of the order in which they appear. For example, $\{1, 2, 3\} = \{3, 2, 1\}$. See also equivalent sets.
equate coefficients	If two polynomials are known to be identical, then the coefficients of like terms must be equal. For example, if $ax^2 + bx + c = px^2 + qx + r$, for all x, then to equate coefficients is to conclude that $a = p$, $b = q$ and $c = r$.
equation	In algebra, a statement that two expressions have the same value for certain values of the variable(s) involved. For example, $x^2 - 3x = 5x - 16$, which is only true if $x = 4$. To solve an equation is to determine the value(s) of the variable that make the statement true. These values are called the roots of the equation. The highest power indicates the number of roots. See also identity.
equation of a circle	In coordinate geometry, a circle equation may be written in the form $x^2 + y^2 + 2gx + 2fy + c = 0$ where the centre of the circle is $(-g, -f)$ and the radius is $\sqrt{g^2 + f^2 - c}$. All points on the circle satisfy the equation.
equation of a circle formula	In coordinate geometry, $(x - h)^2 + (y - k)^2 = r^2$ where (h, k) is the centre and r the radius. The above formula simplifies to $x^2 + y^2 = r^2$ when the centre is $(0, 0)$. The formula can be rearranged to read $x^2 + y^2 + 2gx + 2fy + c = 0$ where the centre of the circle is $(-g, -f)$ and the radius is $\sqrt{g^2 + f^2 - c}$.
equation of a line	In coordinate geometry, the equation of a line may be written in the form $ax + by + c = 0$. All points on the line satisfy the equation.

equation of a line formula

In coordinate geometry, we use the formula $y - y_1 = m(x - x_1)$ to find the equation of a line, where (x_1, y_1) is a point on the line and m is the slope.
The formula may not be used to find the equation of a vertical line, since its slope is undefined.

equiangular triangles

Triangles that contain the same angles.
They are also called similar triangles.

When two triangles are equiangular, corresponding sides are proportional.

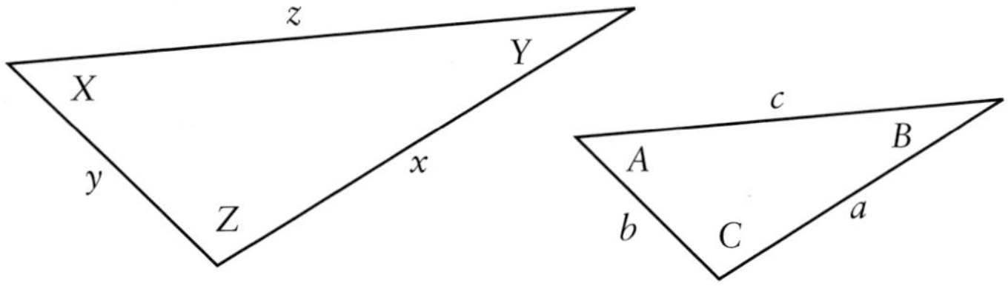

So if $A = X$, $B = Y$ and $C = Z$ then $\frac{a}{x} = \frac{b}{y} = \frac{c}{z}$

See also congruent triangles.

equidistant

When points are the same distance from one or more points or lines they are said to be equidistant from the point(s) or line(s).

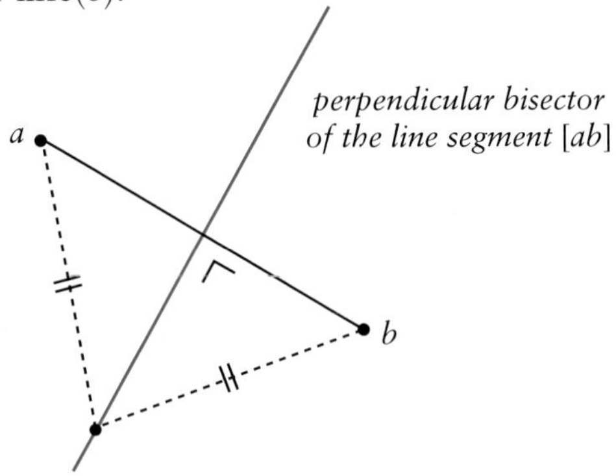

Every point on the perpendicular bisector of a line segment is equidistant from the end points of the line segment.

See also bisector of an angle and mediator of a line segment.

equilateral triangle	A **triangle** containing three **sides** of equal length. The **angles** are also equal, each one being 60°. See also **isosceles triangle** and **scalene triangle**.
equipollent couples	The **couple** (a, b) is equipollent to the couple (c, d) if: 1. $\|ab\| - \|cd\|$ 2. ab is **parallel** to cd 3. ab is in the same direction as cd. A **translation** produces a **set** of equipollent couples.
equivalent fractions	**Fractions** that have the same value. Given any fraction, an equivalent fraction can be created by multiplying or dividing the top and bottom by the same number. For example, $\frac{2}{4}$ is equivalent to $\frac{1}{2}$.
equivalent ratios	**Ratios** that have the same value. Given any ratio, an equivalent ratio can be created by multiplying or dividing each number in the ratio by the same number. For example, $\frac{2}{3}:\frac{3}{4}$ is equivalent to 8:9 (multiplying both numbers by 12).
equivalent sets	**Sets** that contain the same number of **elements**. See also **cardinal number** and **equal sets**.
error	When an approximation to a number is made, the error is obtained by subtracting the approximation (**estimate**) from the exact number (**true value**). Error = True Value – Estimate See also **percentage error**.
estimate	An approximation to a number. For example, 2 is an estimate to 1.76. See also **error** and **percentage error**.

Euclid	Euclid (c. 330–260 BC). Greek mathematician, who wrote what is considered by many to be the most important mathematical text of all time, the *Elements*. It was written about 2300 years ago and has appeared in more editions than any work apart from the Bible. It comprises thirteen books, nine of which deal with plane and solid **geometry** and four with number theory. The commentator Proclus (410–485) wrote that 'The *Elements* bears the same relation to the rest of mathematics as do the letters of the alphabet to language.' Little is known about the life of Euclid other than that he taught at The Museum, a university in Alexandria, Egypt. He is thought to have studied at Plato's Academy.
Euler	Leonhard Euler (1707–83). Born in Basel, Switzerland, Euler graduated with honours from Basel University at the age of fifteen. He hired Johann Bernoulli (see **Bernoulli, Johann**) to give him private tuition in mathematics. By 1750 he was regarded as the foremost mathematician in Europe. He made significant contributions to many areas of mathematics, especially to **geometry**, **trigonometry** and **calculus**. Euler introduced much of the notation that we now take for granted, for example, π, e, Σ, $\log x$, $\sin x$, $\cos x$, $f(x)$. He also introduced the idea of using capital letters for **angles** in trigonometry and lowercase letters for **sides**. Despite becoming almost totally blind, he continued his mathematical work until his death. Because of his amazing memory, he could perform detailed calculations in his head which he then dictated to his sons.

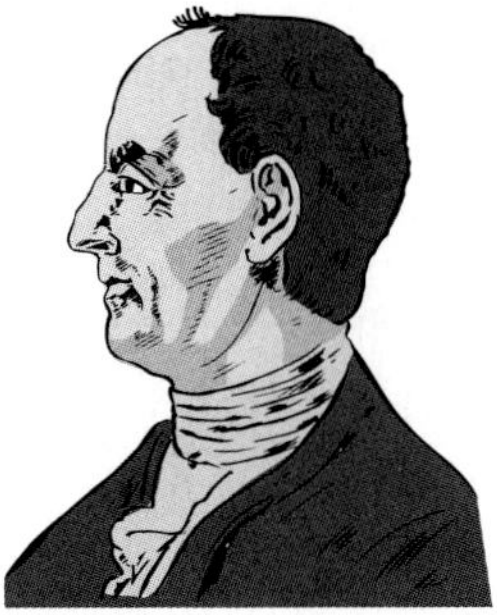

even number	A number that has no remainder when divided by 2. See also prime number and odd number.
event	The name given to a particular outcome or set of outcomes when an experiment is conducted in probability. See also trial.
expand	To take an expression involving an index or a product and rewrite it in another form, either by multiplication or using the binomial theorem. For example, $(x - y)^3$ expands to give $x^3 - 3x^2y + 3xy^2 - y^3$. See also expansion.
expansion	In algebra, when a bracket containing a sum or difference is raised to a power, it is said to be expanded. For example, $(2x + 1)^2 = 4x^2 + 4x + 1$. See also binomial theorem.
experiment	In probability, an action that has a number of possible outcomes. For example, rolling a die or choosing a card from a deck.
explicit differentiation	Differentiation where one variable is expressed in terms of the other. For example: Find $\frac{dy}{dx}$ when $y = \sqrt{x^2 - 1}$. See also implicit differentiation.
exponential function	$e^x = 1 + x + \frac{x^2}{2!} + \frac{x^3}{3!} + \frac{x^4}{4!} + \ldots\ldots\ldots\ldots$ where $e = \lim_{n \to \infty}\left(1 + \frac{1}{n}\right)^n \approx 2.71828$ This expansion of e^x is created using the Maclaurin series formula. The significance of the exponential function is in differentiation, where it is the only function whose derivative is itself, i.e. $\frac{d}{dx}(e^x) = e^x$.

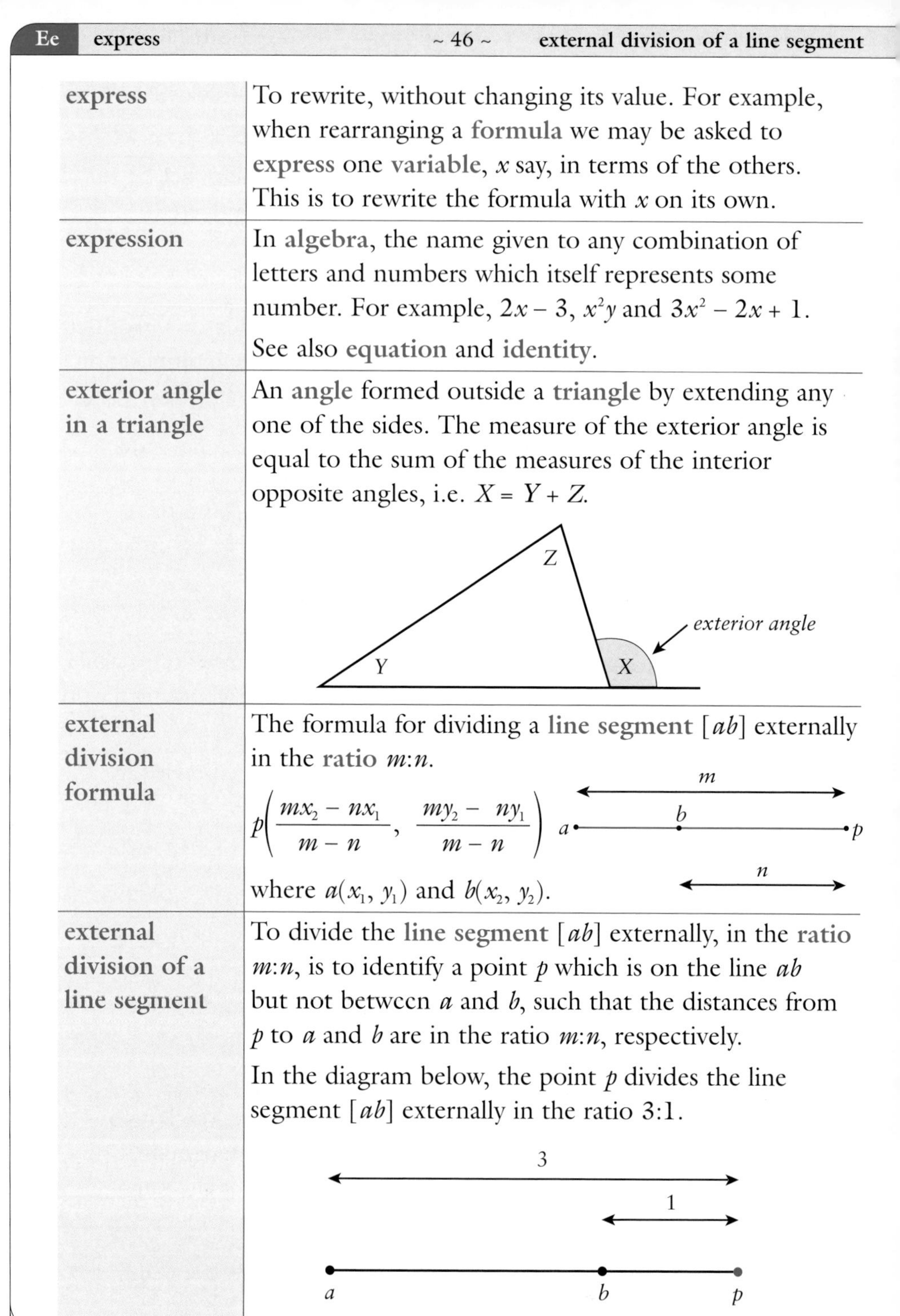

express

To rewrite, without changing its value. For example, when rearranging a **formula** we may be asked to **express** one **variable**, x say, in terms of the others. This is to rewrite the formula with x on its own.

expression

In **algebra**, the name given to any combination of letters and numbers which itself represents some number. For example, $2x - 3$, x^2y and $3x^2 - 2x + 1$.

See also **equation** and **identity**.

exterior angle in a triangle

An **angle** formed outside a **triangle** by extending any one of the sides. The measure of the exterior angle is equal to the sum of the measures of the interior opposite angles, i.e. $X = Y + Z$.

external division formula

The formula for dividing a **line segment** [ab] externally in the **ratio** m:n.

$$p\left(\frac{mx_2 - nx_1}{m - n}, \frac{my_2 - ny_1}{m - n}\right)$$

where $a(x_1, y_1)$ and $b(x_2, y_2)$.

external division of a line segment

To divide the **line segment** [ab] externally, in the **ratio** m:n, is to identify a point p which is on the line ab but not between a and b, such that the distances from p to a and b are in the ratio m:n, respectively.

In the diagram below, the point p divides the line segment [ab] externally in the ratio 3:1.

extraneous roots	Roots that arise in the course of solving an equation that are not roots of the original equation. They often arise when solving surd equations or equations involving rational functions with the variable on the bottom.

Ff

factor	A number p is a factor of another number q if p divides evenly into q. For example, 3 is a factor of 6. See also highest common factor and multiple.
factor theorem	Given an expression $f(x)$ and a number a, if $f(a) = 0$ then $x - a$ is a factor of $f(x)$. For example, if $f(x) = x^3 - 2x^2 - 4x + 5$, then since $f(1) = 0$, it follows from the factor theorem that $x - 1$ is a factor of $f(x)$. We use the factor theorem to factorise cubic expressions.
factorial	Factorial n is written: $n!$ $n! = n(n-1)!$, where n is a natural number and $0! = 1$. For example, $3! = 3.2!$ $= 3.2.1!$ $= 3.2.1.0!$ $= 3.2.1.1 = 6$ The number of permutations of n different objects when all objects appear in each permutation is $n!$.
factorise	Find factors of. Note: When we factorise in algebra, we convert an expression that is a sum into another expression that is a product, but both expressions have exactly the same value for all values of the variables involved. For example: The factors of $x^2 - y^2$ are $(x + y)(x - y)$ The factors of $3x^2y - 12xy^2$ are $3xy(x - 4y)$ The factors of $x^2 - 3x - 10$ are $(x - 5)(x + 2)$

finite series	A series containing a fixed number of terms.
first derivative test	In differentiation, a method used to identify the nature of a stationary point. Two numbers, one to the left and one to the right of the stationary point, are substituted into the first derivative, $f'(x)$. The resulting signs indicate whether the stationary point is a local maximum, a local minimum or a saddle point. Negative followed by positive $\Rightarrow$ local minimum. Positive followed by negative $\Rightarrow$ local maximum. No change in sign $\Rightarrow$ saddle point. See also second derivative test.
first principles	In differentiation, to use first principles is to calculate $\lim\limits_{h \to 0} \frac{f(x+h) - f(x)}{h}$ step by step. See also derivative.
formula	A set of symbols and numbers that expresses a fact or rule in the form of an equation. For example, $x = \frac{-b \pm \sqrt{b^2 - 4ac}}{2a}$ is the formula used to solve the quadratic equation $ax^2 + bx + c = 0$. The variable that appears on its own in a formula is called the subject of the formula. In the example above, x is the subject.
fraction	A ratio of two whole numbers, $a{:}b$, written in the form $\frac{a}{b}$. The bottom number is called the denominator. The top number is called the numerator. The bottom number must never be 0. See also rational number.

frequency — In **statistics**, how often a particular statistic occurs. In the list of numbers: 1, 3, 4, 1, 2, 4, 1, 5, 1, the number 1 has a frequency of 4.

See also **frequency distribution table**.

frequency distribution table — In **statistics**, a table that tells us how many times (**frequency**) individual items occur when a survey is conducted. For example, the frequency table below shows that in a certain group of people, 4 are aged 12 and 7 are aged 13.

Age	12	13	14	15	16	17
Frequency	4	7	2	6	2	1

See also **grouped frequency distribution table**.

function — A **relation** in which no two **couples** have the same first component and every **element** of the **domain** has an **image**. For example, $\{(a, 1), (b, 3), (c, 3)\}$ is a function.

A function of x, usually denoted $f(x)$, is an **expression** whose value depends only on the value of x, for example, $f(x) = x^2 - 3x + 1$.

The function gives rise to a set of couples of the form $(x, f(x))$, which enables us to draw a **graph** of the function.

Gg

g — Abbreviation for **gram**.

Gauss — Carl Friedrich Gauss (1777–1855). German mathematician who introduced the concept of **complex numbers**. Because of his many contributions to a variety of areas in mathematics and physics, he is regarded as one of the most influential of all mathematicians.

general polar form	In complex numbers, to convert from polar form to general polar form, add $2n\pi$ or $360n°$ to the argument. The number $2(\cos\pi + i\sin\pi)$ is in polar form. The number $2[\cos(\pi + 2n\pi) + i\sin(\pi + 2n\pi)]$ is in general polar form.
general term	In the binomial expansion of $(x + y)^n$, the general term is $$T_{r+1} = \binom{n}{r} x^{n-r} y^r$$ This is used to identify individual terms without having to list the entire expansion. See also binomial theorem.
generator	An element of a group is a generator if every other element may be expressed as a power of that element. For example, i is a generator of the group $\{1, -1, i, -i\}, \times$, since $i^1 = i$, $i^2 = -1$, $i^3 = -i$, $i^4 = 1$. Any element of a finite group whose order is equal to that of the group is a generator of the group.
geometric mean	The nth root of the product of n numbers. For example, the geometric mean of 2, 3 and 4 is $\sqrt[3]{2 \times 3 \times 4} = \sqrt[3]{24}$. See also arithmetic mean.
geometric sequence/series	A sequence or series in which the ratio of consecutive terms is constant. For example, $3 + 1 + \frac{1}{3} + \frac{1}{9} + \frac{1}{27} + \ldots\ldots$ is a geometric series with common ratio $\frac{1}{3}$. See also arithmetic sequence/series and arithmetico-geometric series.

geometry

A branch of mathematics in which we study objects that can be drawn on a **two-dimensional** plane surface to discover their properties. For example, we discover that the opposite sides of every parallelogram are equal in length.

Geometry splits into two parts: Euclidean geometry, in which we study the plane with the help of **theorems**, and **coordinate geometry**, where we use **algebra**.

given

A result that is known to be true. Usually stated at the beginning of the **proof** of a **theorem**.

gradient

See **slope**.

gram

Unit of mass. Abbreviated to g.

1 **kilogram** (kg) = 1000 grams (g)

1g = 1000 **milligrams** (mg)

graph

A diagram showing the relationship between quantities. Below is part of the graph of the quadratic function $f(x) = x^2 - x - 6$.

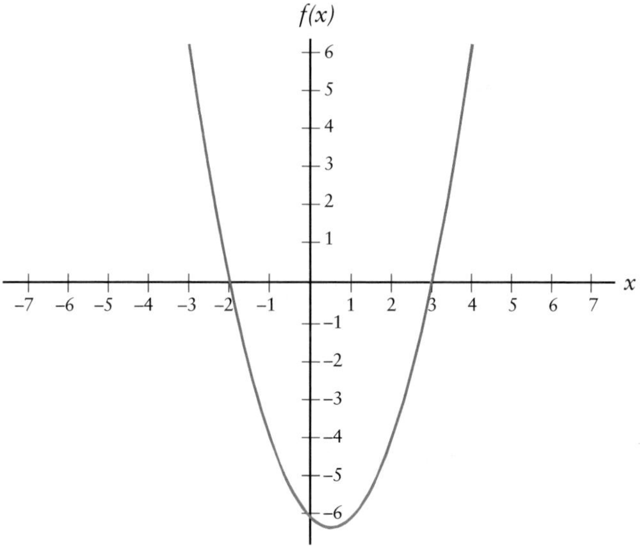

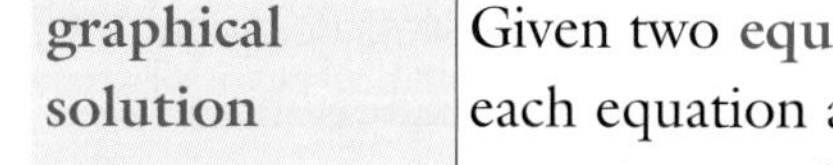

graphical solution	Given two **equations**, we can draw a **graph** to represent each equation and identify the **solution**(s) to the **equations** as the **point**(s) of **intersection** of the graphs.

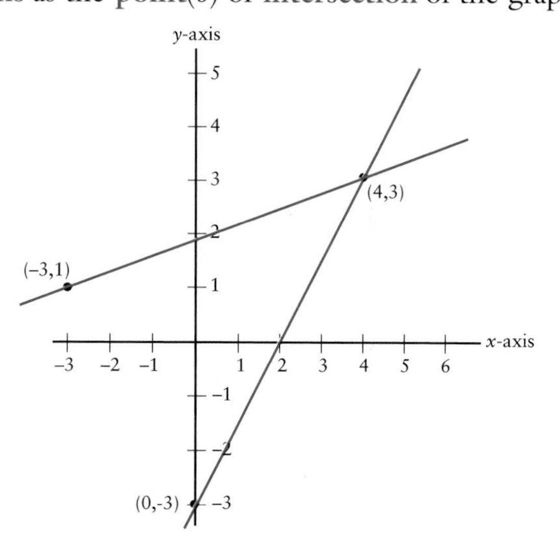

This **diagram** shows that $x = 4$, $y = 3$ is the solution to the **simultaneous equations** $2x - 7y + 13 = 0$ and $3x - 2y - 6 = 0$.

gross income	Income before any deductions have been made. See also **net income**, **income tax** and **tax credits**.
gross tax	When **gross income** is less than the **standard rate cut-off point**, gross tax is a percentage of gross income. When the gross income is greater than the standard rate cut-off point, the gross income is split into two parts in order to calculate the gross tax, for example: A man has a gross income of €60,000. The standard rate cut-off point is €25,000. The standard rate of tax is 20% and the higher rate is 40%. His gross tax is calculated as follows: €25,000 @ 20% = €5,000 €35,000 @ 40% = €14,000 Gross Tax = €19,000 **Tax credits** are then subtracted from the gross tax to calculate the tax to be paid.

group

A set with a binary operation that satisfies four conditions:

1. The operation is associative
2. The set is closed under the operation
3. There is a unique identity element
4. Every element has an inverse element

See also abelian group.

grouped frequency distribution table

In statistics, a table that tells us how many times (frequency) groups of items occur when a survey is conducted. For example, this table indicates that in a certain group of people, 7 are aged between 5 and 7 and 2 are aged between 7 and 9.

Age	5–7	7–9	9–11	11–13	13–15
Frequency	7	2	4	2	3

See also frequency distribution table and cumulative frequency table.

Hh

half-angle formula

A formula in trigonometry that expresses the value of a function of an angle in terms of functions of twice the angle.

For example, $\cos^2 A = \frac{1}{2}(1 + \cos 2A)$

See also double-angle formula.

half-plane

All points on one or other side of a line.

A half-plane is represented by an inequality in the form $ax + by > c$ or $ax + by < c$ where $ax + by = c$ is the equation of the line.

The shaded region in the diagram represents the half-plane $3x - 2y - 6 < 0$.

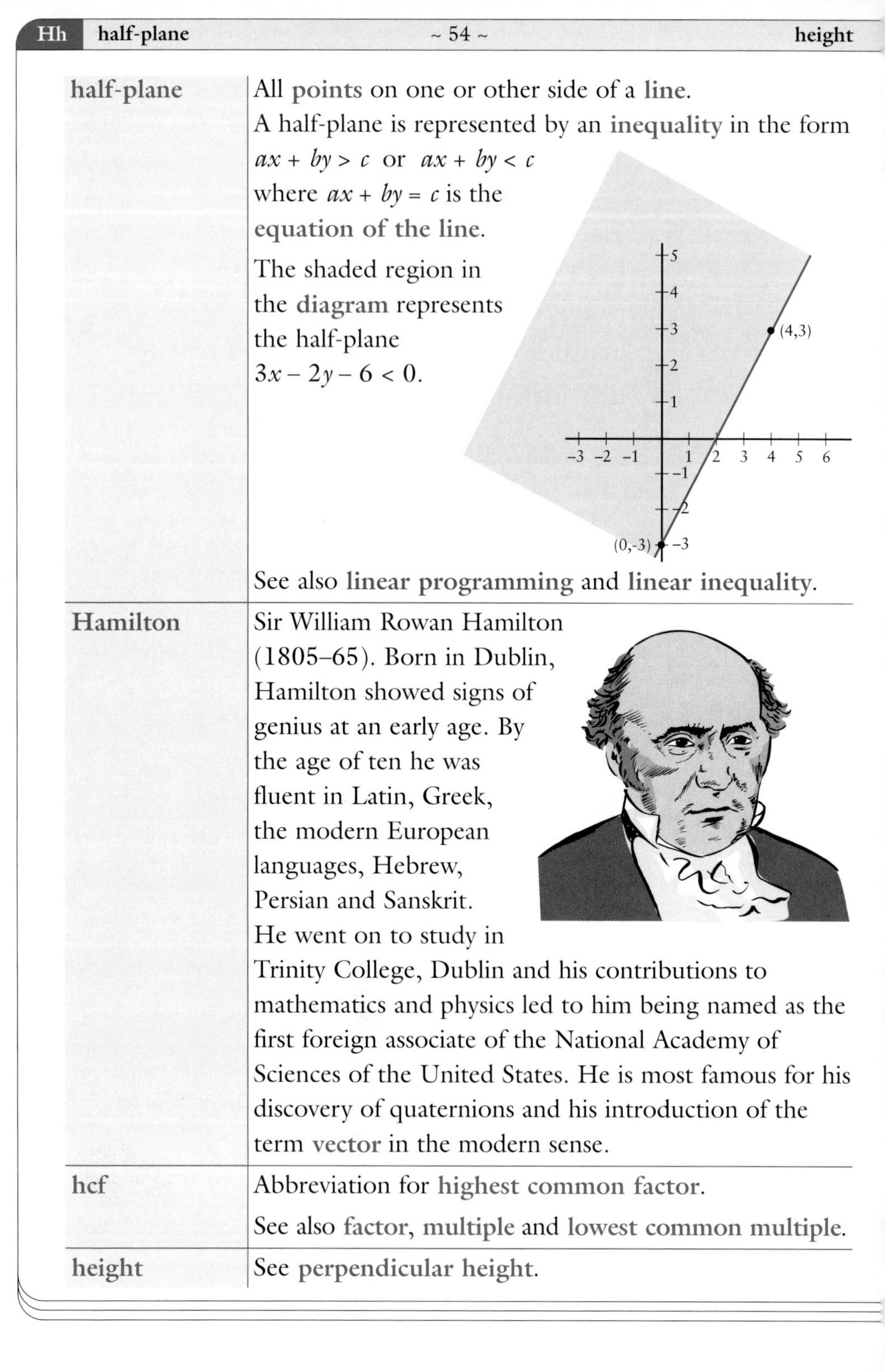

See also linear programming and linear inequality.

Hamilton

Sir William Rowan Hamilton (1805–65). Born in Dublin, Hamilton showed signs of genius at an early age. By the age of ten he was fluent in Latin, Greek, the modern European languages, Hebrew, Persian and Sanskrit. He went on to study in Trinity College, Dublin and his contributions to mathematics and physics led to him being named as the first foreign associate of the National Academy of Sciences of the United States. He is most famous for his discovery of quaternions and his introduction of the term vector in the modern sense.

hcf

Abbreviation for highest common factor.

See also factor, multiple and lowest common multiple.

height

See perpendicular height.

hemisphere

Half of a sphere.

Its volume is $\frac{2}{3}\pi r^3$

Its surface area is $3\pi r^2$

See also cylinder and cone.

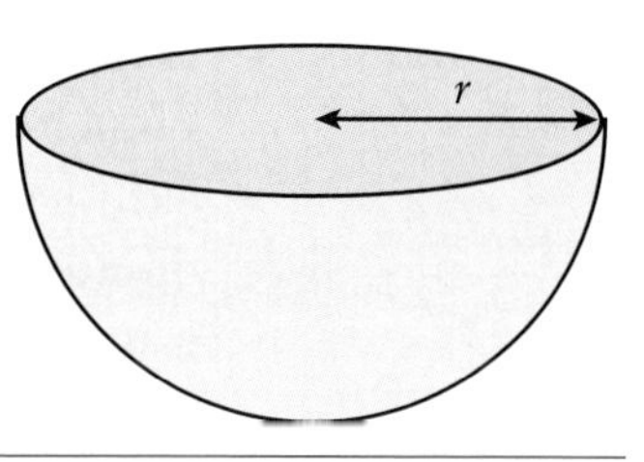

hexagon

A plane figure with six sides.
A regular hexagon has sides of equal length. The angle between adjacent sides is 120°.

See also pentagon.

highest common factor

The largest number that divides evenly into two or more numbers. For example, 1, 2 and 4 are common factors of 8 and 12. The highest common factor is 4. Abbreviated to hcf.

See also factor, multiple and lowest common multiple.

histogram

A graph in statistics. Rectangular blocks, possibly of different widths, are drawn with no space between them, so that the area of each block represents the frequency over that interval in a grouped frequency distribution table.

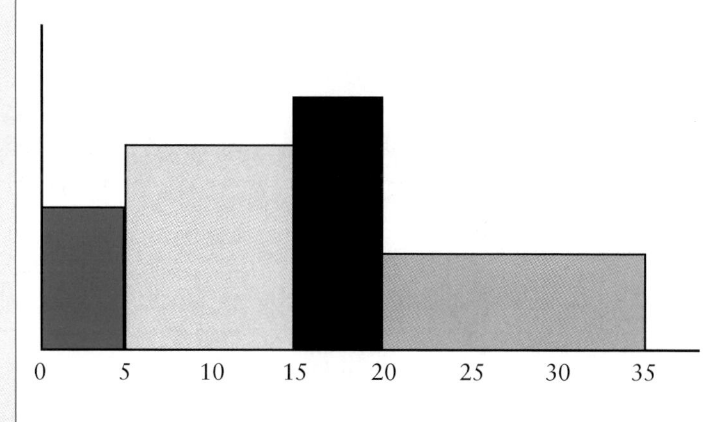

See also bar chart, pie chart and trend graph.

l'Hôpital	Guillaume Francois Antoine de l'Hôpital (1661–1704). French mathematician who was the author of the first textbook on differential calculus. After his death it was discovered that another mathematician, Johann Bernoulli (see **Bernoulli, Johann**), was responsible for many of the ideas in the text. Bernoulli had been paid by l'Hôpital to teach him calculus.
l'Hôpital's rule	A rule to evaluate **limits** that have produced the **indeterminate** form $\frac{0}{0}$. $$\lim_{x \to a} \frac{f(x)}{g(x)} = \frac{f'(a)}{g'(a)}$$ For example, $$\lim_{x \to 2} \frac{x^2 - 4}{x - 2} = \frac{4 - 4}{2 - 2} = \frac{0}{0}$$ so we apply l'Hôpital's rule $$\lim_{x \to 2} \frac{x^2 - 4}{x - 2} = \lim_{x \to 2} \frac{2x}{1} = 4$$
horizontal	Straight across from left to right. See also **vertical**.
hypotenuse	The longest side in a **right-angled triangle**. It is always opposite the **right angle**. hypotenuse See also **Pythagoras' theorem**.
hypothesis	An unproved theory.
hypothesis testing	In **probability**, the determination as to the likelihood that a particular outcome of an **experiment** is 'unusual.' What we consider to be unusual is determined by the **significance level** that is given in the question. For instance, at the 5% level of significance, an unusual result is one which has a **probability** of 0.05 or less.

Ii

identity	In **algebra** or **trigonometry**, a statement that two **expressions** are identical, although they look different, for all values of the **variables** involved. For example, $x^2 + y^2$ is identical to $(x + y)^2 - 2xy$, for all values of x and y. $\sin 2A$ is identical to $2\sin A\cos A$, for all values of A. We may be required to prove an identity. To do this, take one side (usually the more complicated) and **simplify** it to end up with the other side. See also **equation**.
identity couple	A **couple** in which both components are the same. For example, (a, a).
identity element	An element which, when combined with another under an operation, does not change it. For example, 1 is the identity element for multiplication since $a \times 1 = a$, for all a. 0 is the identity element for addition since $a + 0 = a$, for all a. See also **group** and **inverse element**.
identity map	The name given to a **transformation** in which a set of **points** is mapped onto itself.
identity matrix	A **matrix** which, when combined with another, does not change it. The most important identity matrix is the one for multiplication of 2×2 matrices: $\begin{pmatrix} 1 & 0 \\ 0 & 1 \end{pmatrix}$
image	When a **transformation** is applied to an object, the resulting object is referred to as the image. When a **linear transformation** is performed, the image is denoted (x', y').

imaginary axis	In complex numbers, the vertical axis in an Argand diagram. It is used to illustrate imaginary numbers.
imaginary numbers	Numbers involving the square root of a negative number. For example, i, $2i$, $i\sqrt{3}$, where $i = \sqrt{-1}$. See also real numbers, natural numbers, rational numbers and integer.
implicit differentiation	A form of differentiation where we are given an equation involving x and y in which one variable is not written in terms of the other, to find $\frac{dy}{dx}$. For example: Find $\frac{dy}{dx}$ when $x^2 - 3xy + y^2 = 1$ The key to implicit differentiation is that $\frac{d}{dx}f(y) = \frac{d}{dy}f(y) \times \frac{dy}{dx}$ Therefore, $\frac{d}{dx}\left(y^2\right) = 2y\frac{dy}{dx}$ See also explicit differentiation.
improper fraction	A fraction in which the top number (numerator) is bigger than the bottom number (denominator). It is also called a top-heavy fraction. Every improper fraction can also be written as a mixed number, for example, $\frac{7}{3} = 2\frac{1}{3}$ See also proper fraction.
improper subset	Every set has two improper subsets, the set itself and the empty set. Therefore, the only improper subsets of the set $\{a, b, c\}$ are the set itself, $\{a, b, c\}$, and the empty set, $\{\ \}$. See also proper subset.

incentre of a triangle	The point of intersection of the bisectors of the angles in a triangle. It is the centre of the incircle. 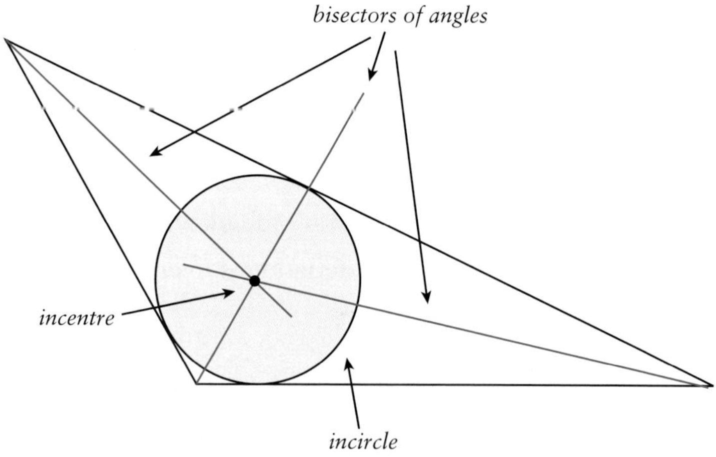 See also centroid, circumcentre and orthocentre.
incircle	A circle inside a triangle that touches the three sides (see diagram at incentre of a triangle). The three sides of the triangle are tangents to the incircle. See also incentre of a triangle, circumcircle and circumcentre.
income tax	A percentage of a person's gross income less his/her tax credits, which is paid to the government. See also net income.
increasing function	A function $f(x)$ is said to be increasing if $f(a) > f(b)$ for all $a > b$. When a function is increasing, its derivative is positive. See also decreasing function.
indefinite integral	An integral in which the answer is not definite since it contains an unknown constant. Always include the constant C in the answer. For example, $\int x^2\, dx = \frac{1}{3}x^3 + C$ See also definite integral.

independent events	In probability, events are independent if the probability that one will occur is not influenced by whether or not the other has already occurred. Mathematically, E and F are independent if the probability that E will occur is equal to the probability that E will occur, given that F has already occurred, i.e. $P(E) = P(E \mid F)$ If E and F are independent, $P(E \cap F) = P(E).P(F)$ See also dependent events.
indeterminate	When evaluating a limit, the result $\frac{0}{0}$ is indeterminate. This means that a limit may or may not exist and that some more work is required before a conclusion may be reached.
index	When a number is in the form a^x, x is called the index. Logarithm is another word for index. See also power.
index equation	An equation where the variable is in the index. For example, $2^{3x-1} = 32$.
index notation	See scientific notation.
indices	Plural of index.
induction	A method of proof that is only suitable for certain problems involving natural numbers. There are four stages to the proof: 1. Demonstrate that the result is true for the smallest number in the set for which proof is required 2. Assume that the result is true for some number, k 3. Prove that if the result is true for k, then it must also be true for the next whole number, $k + 1$ 4. Conclude that the result must, therefore, be true for all numbers in the required set See also deduction.

inequality	A statement in **algebra** that one **expression** is greater than another. It will involve one of the following symbols: $<$, $\leq$, $>$ or $\geq$. Some inequalities must be solved, in which case we find the values of the **variable** that will make this statement true. Others must be proved, in which case we prove that the statement is true for all values of the variable.
infinite	Having size or length greater than any **whole number**.
infinite series	A **series** containing an unlimited number of terms. All infinite series are either **convergent** or **divergent**.
infinity	Mathematical quantity larger than any number. The symbol for infinity is ∞.
inflection point	A **point** on a curve that satisfies two conditions: 1. $\frac{d^2y}{dx^2} = 0$. 2. $\frac{d^2y}{dx^2}$ has different signs on either side of the point. 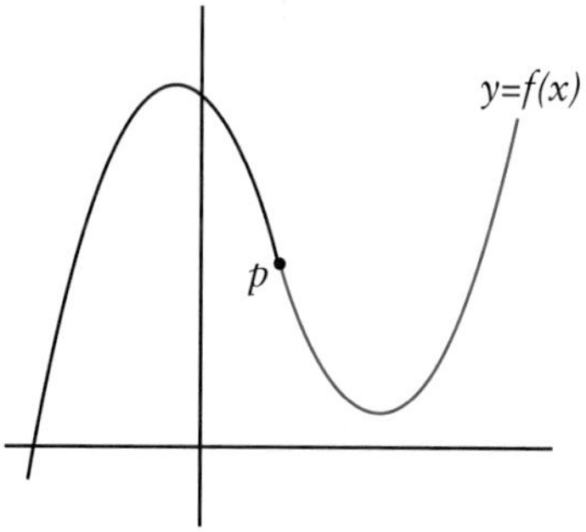 On the black section of the graph, $\frac{dy}{dx}$ is decreasing and $\frac{d^2y}{dx^2}$ is negative. At p, $\frac{dy}{dx}$ has reached its minimum value and so $\frac{d^2y}{dx^2}$ is 0. On the blue section, $\frac{dy}{dx}$ is increasing and $\frac{d^2y}{dx^2}$ is positive. p is an inflection point. See also **saddle point** and **stationary point**.
information	In **statistics**, **data** to which a meaning has been given.

integer	Any element of the set {.......... –4, –3, –2, –1, 0, 1, 2, 3, 4, ...}. Integers are also called **whole numbers** and are represented by the symbol Z. See also **natural numbers**, **rational numbers** and **real numbers**.
integration	The reverse of **differentiation**. It is used to find **area** and **volume** of irregularly shaped objects. The symbol for integration is $\int$. See also **definite integral**, **indefinite integral** and **integration by parts**.
integration by parts	A method for integrating certain products, which involves the formula $\int u.dv = uv - \int v.du$ See also **integration**, **definite integral** and **indefinite integral**.
interest	When a sum of money is deposited by a person in a bank (or other financial institution), interest is paid to that person, so long as his/her money remains in the bank. It is a **percentage** of the money held and is usually paid at regular intervals. Or Money that must be repaid when borrowing, in addition to the sum borrowed. See also **amount**, **compound interest** and **principal**.
interior angles	When **parallel lines** are cut by a **transversal**, interior angles appear between the parallel lines on the same side of the transversal. They add up to 180°. 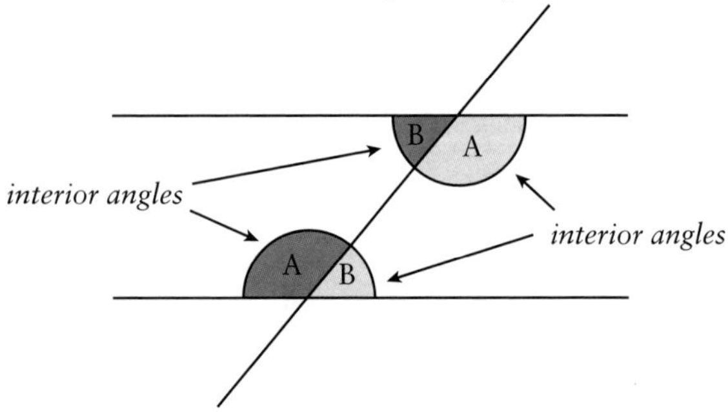

interior angles in a triangle
The angles inside a triangle. They add up to 180°.

internal division formula
The formula

$$p\left(\frac{mx_1 + nx_2}{m + n}, \frac{my_1 + ny_2}{m + n}\right)$$

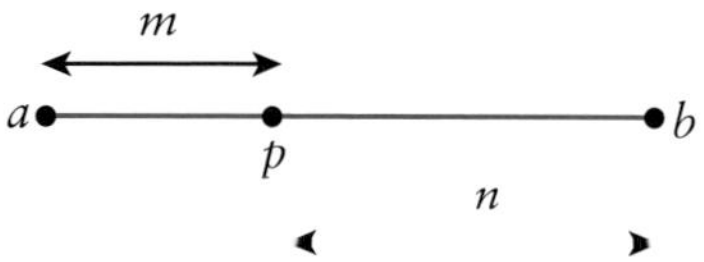

gives the coordinates of the point, p, which divides the line segment $[ab]$ internally in the ratio $m:n$, where $b\,(x_1, y_1)$ and $a(x_2, y_2)$.

See also midpoint formula and external division formula.

internal division of a line segment
To divide the line segment $[ab]$ internally in the ratio $m:n$, is to identify a point p which is on the line segment $[ab]$ such that the distances from p to a and b are in the ratio $m:n$, respectively.

In the diagram below, the point p divides the line segment $[ab]$ internally in the ratio 2:1.

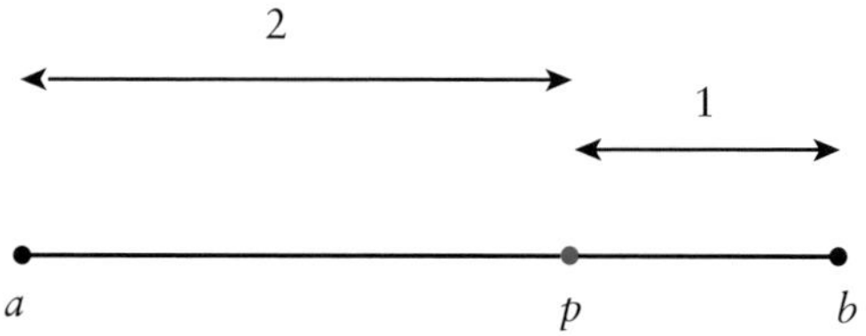

See also internal division formula.

interquartile range
In statistics, a number found from a cumulative frequency curve by subtracting the lower quartile from the upper quartile. It is a measure of the dispersion of the data. In this example, the interquartile range is $27 - 15 = 12$.

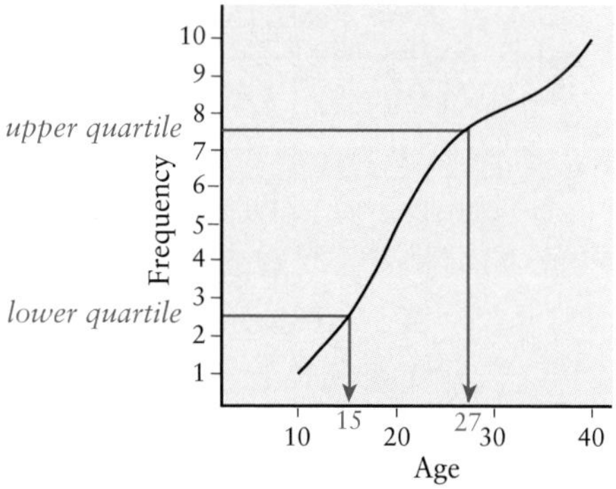

intersect	To cut across or touch. **Parallel** lines never intersect. A **tangent** intersects a **circle** at only one **point**.
intersection	The **point** of intersection of two **lines** is the point where they cross. It is the only point that is on both lines. When curves intersect lines or other curves, there may be more than one point of intersection. The coordinates of the point of intersection of two curves can be obtained from their equations using a **graph** or using a method referred to as **simultaneous equations**.
intersection of two sets	The **set** of **elements** that are common to the sets. The symbol for intersection of sets is ∩. For example, if A = {1, 2, 3, 4} and B = {2, 4, 6, 8} then A ∩ B {2, 4}. On a **Venn diagram**, the intersection is represented by the region where the ovals overlap. The diagram shows $A \cap B \cap C$. See also **union of sets** and **set difference**. A B C
interval	The **set** of numbers between two other numbers. For example, $-1 < x < 2$, $x \in R$ describes all **real numbers** between –1 and 2, exclusive. $-1 \leq x \leq 2$, $x \in R$ describes all real numbers between –1 and 2, inclusive.
interval of convergence	The **set** of numbers for which a **power series** converges. It can be determined using the **ratio test**.
invariant	Does not change. When dealing with **linear transformations**, you may be asked to examine the **area** of a **triangle** before and after a transformation, to determine whether or not **area** is invariant under a linear transformation.

inverse element	In a group, with a binary operation $*$, every element a has an inverse element a^{-1}, such that $a * a^{-1} = e = a^{-1} * a$, where e is the identity element.
inverse function	If f is a function, then its inverse is written f^{-1}. It is formed by reversing all couples of f. Therefore, if (a, b) is a couple of the function f, then (b, a) will be a couple of the inverse function. Any function followed by its inverse produces the identity map, i.e. $f^{-1} \circ f(x) = x$. The inverse of the function $\sin x$ is $\sin^{-1} x$. The inverse of the function e^x is $\log_e x$.
inverse of a 2 × 2 matrix	The inverse of $\begin{pmatrix} a & b \\ c & d \end{pmatrix}$ is $\dfrac{1}{ad - bc}\begin{pmatrix} d & -b \\ -c & a \end{pmatrix}$ When a 2 × 2 matrix is multiplied by its inverse, the result is the identity matrix $\begin{pmatrix} 1 & 0 \\ 0 & 1 \end{pmatrix}$. If $ad - bc = 0$, the matrix does not have an inverse and is called a singular matrix.
inverse proportion	A process where one quantity decreases while another increases, their product remaining constant. For example, if three men complete a job in six days, then two men working at the same rate would take nine days. Notice that the number of men has decreased and the number of days has increased, but the product in both cases is eighteen. See also inversely proportional and direct proportion.
inverse trigonometric function	$\sin^{-1} x$ is an angle between $-\dfrac{\pi}{2}$ and $\dfrac{\pi}{2}$, inclusive, whose sin is x. $\tan^{-1} x$ is an angle between $-\dfrac{\pi}{2}$ and $\dfrac{\pi}{2}$, exclusive, whose tan is x.
inversely proportional	Two numbers x and y are inversely proportional to two others a and b if $\dfrac{x}{y} = \dfrac{b}{a}$.

irrational numbers	Numbers that cannot be expressed in the form $\frac{a}{b}$, where a and b are whole numbers and b is not 0. For example, $\sqrt{2}$ $\sqrt{12}$, e π. Note: A decimal number is irrational if it goes on forever without recurring; otherwise it is rational. See also integer, natural numbers, real numbers and complex numbers.
isometry	A measure conserving transformation. It conserves length, area and measure of angle. Translations, axial symmetries, central symmetries and rotations are all isometries. Parallel projections and some linear transformations are not.
isomorphic groups	Two groups, A, ⊗ and B, * are isomorphic if a bijection f exists from A to B such that $f(x \otimes y) = f(x)^* f(y)$, for all $x, y \in$ A. See also abelian group.
isomorphism	A one-to-one correspondence between the elements of two isomorphic groups.
isosceles triangle	A triangle containing two sides of equal length. The angles opposite these sides are also equal. See also equilateral triangle and scalene triangle. 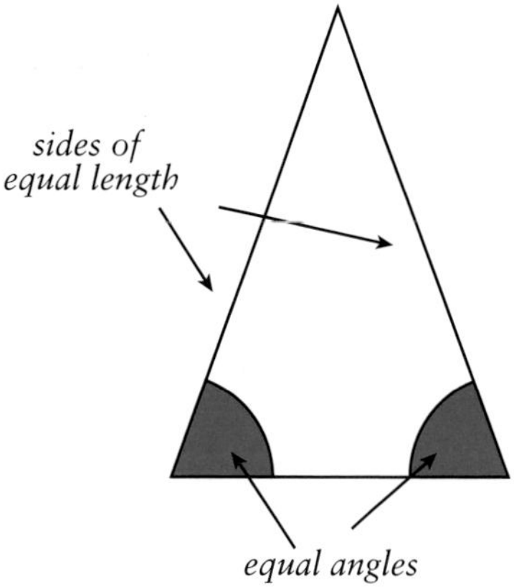
iteration	Each step in an iterative procedure.

iterative procedure	A procedure in which the output of a **formula** becomes the input when the formula is reapplied. An example of an iterative procedure is the method used when applying the **Newton-Raphson formula** to approximate a **root** of an **equation**.

Kk

kg	Abbreviation and symbol for **kilogram**.
kilogram	Unit of mass. Abbreviated to kg. 1 kilogram (kg) = 1000 **grams** (g) 1g = 1000 **milligrams** (mg)
kilometre	Unit of measurement. Abbreviated to km. 1 kilometre (km) = 1000 **metres** (m) 1m = 100 **centimetres** (cm) 1cm = 10 **millimetres** (mm)
km	Abbreviation for **kilometre**.

Ll

l	Abbreviation for **litre**.
Lagrange	Joseph Louis, Comte de Lagrange (1736–1813). Italian mathematician who made significant contributions to many branches of mathematics, including probability theory, theory of equations and group theory. In 1788, he published his most important work, *Analytical Mechanics*, in which he extended the work of **Newton**, the **Bernoulli** family and **Euler**. Lagrange introduced the notation $f'(x)$ to represent the **derivative**.

Lagrange's theorem	The order of every subgroup of a group is a factor of the order of the group.
lcm	Abbreviation for lowest common multiple. See also highest common factor.
Leibniz	Gottfried Wilhelm Leibniz (1646–1716). German mathematician who developed calculus independently of Newton.
length	The length of a line segment $[ab]$ is a measure of the distance from a to b and is written $\|ab\|$. The length of an arc ab is a measure of the distance around the circumference of the circle from a to b. The most common units of length are millimetre (mm), centimetre (cm), metre (m) and kilometre (km). See also perimeter.
l'Hôpital	See l'Hôpital under H.
l'Hôpital's rule	See l'Hôpital's rule under H.
like terms	In algebra, terms that contain the same variables and the same index on each variable. For example, in the expression $x^2 - 4x^2y + 3x^2y - y^2$, $-4x^2y$ and $3x^2y$ are like terms. We may add like terms by adding their coefficients, but we must not change the variables or their indices: $-5x^2y + 3x^2y = -2x^2y$. Terms that are not alike are called unlike terms.
limit	1. $\lim_{n \to \infty} f(n) = L$ means that the value of $f(n)$ can be made as close to L as we choose by replacing n with a big enough number. However, no matter how big n becomes, $f(n)$ will never actually equal L. For example, when n is very big, $\dfrac{n}{n+2}$ will be very close to 1. However, no matter how big n is, the fraction will never equal 1 since the top and bottom will always be different.

limit	2. $\lim\limits_{n \to a} f(n) = L$ means that the value of $f(n)$ can be made very close to L by replacing n with a value very close to a. The value of $f(n)$ can be made as close to L as we choose by substituting a value for n that is very close to a, but it can never be made to equal L. For example, $\lim\limits_{n \to 2} \frac{n^2 - 4}{n - 2} = 4$, so $\frac{n^2 - 4}{n - 2}$ will be very close to 4 when n is very close to 2. See **l'Hôpital's rule** under H.
limits of an integral	The numbers that appear at the top and bottom of a **definite integral**, which are to be substituted for the **variable**, having integrated. For example, $\int_0^1 x^2 dx = \left[\frac{1}{3}x^3\right]_0^1 = \frac{1}{3}$
line	A line ab is an **infinite** straight line that passes through the points a and b. A line has no thickness, so only one line can be drawn that passes through both a and b. a b See also **line segment**.
line of vision	If you stand at a point a and look at another point, b, the line along which you look is called the line of vision. See also **angle of elevation** and **angle of depression**.
line segment	The line segment $[ab]$ is a straight line joining the point a to b, but not extending beyond a or b. a b See also **line**.

linear combination	A linear combination of the **vectors** $\vec{x}$ and $\vec{y}$ is in the form $a\vec{x} + b\vec{y}$ where a and b are **scalars**.
linear equation	In **algebra**, an **equation** that does not contain **powers** or **products** of the **variables** involved. $3(2x - 1) - (x - 2) = \frac{5x}{2}$ is a linear equation in one variable. $3x - 7y = 1$ is a linear equation in two variables. If the **couples** that satisfy a linear equation in two variables are plotted as **points** on a **graph**, the points are always in a straight line. See also **quadratic equation** and **cubic equation**.
linear inequality	An inequality involving x and/or y that has no indices on x or y. It can be represented by a half-plane in **coordinate geometry**.
linear programming	The use of **linear inequalities** to solve a certain type of problem involving **constraints**.
linear transformation	If $\vec{z} = a\vec{x} + b\vec{y}$ then the **vector** $\vec{z}$ is said to be a **linear combination** of $\vec{x}$ and $\vec{y}$. $f(\vec{x})$, $f(\vec{y})$ and $f(\vec{z})$ are the images of $\vec{x}$, $\vec{y}$ and $\vec{z}$ under some transformation, f. The transformation f is linear if: $f(\vec{z}) = a.f(\vec{x}) + b.f(\vec{y})$. The linear transformations on the Leaving Certificate Higher Level course will be presented in the form: $f:(x, y) \rightarrow (x', y')$, where $x' = ax + by$ $y' = cx + dy, \quad ad - bc \neq 0$ (x, y) represents a point and (x', y') is its image under the transformation.

litre	Unit of volume. Abbreviated to l. 1 litre = 1000cm^3 $1l$ = 1000 millilitres (ml)
local maximum point	A point on a curve where 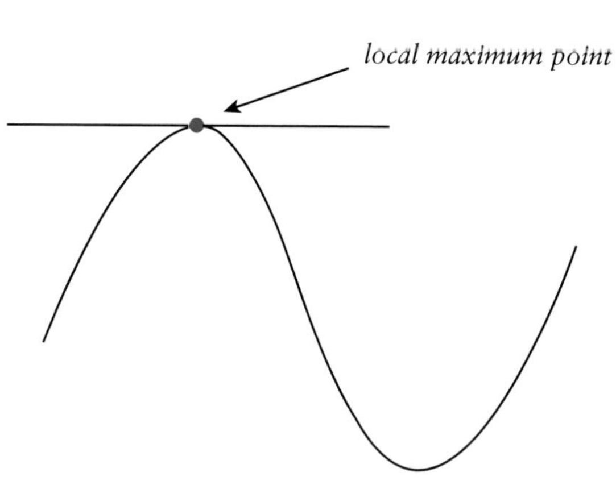 1. $\frac{dy}{dx} = 0$. 2. $\frac{dy}{dx}$ is positive just to the left of the point and negative just to the right of it. Or $\frac{d^2y}{dx^2} < 0$ at the local maximum. It is a turning point. It belongs to a category of points called stationary points, at which $\frac{dy}{dx} = 0$. See also local minimum point and point of inflection.

local minimum point	A point on a curve where: 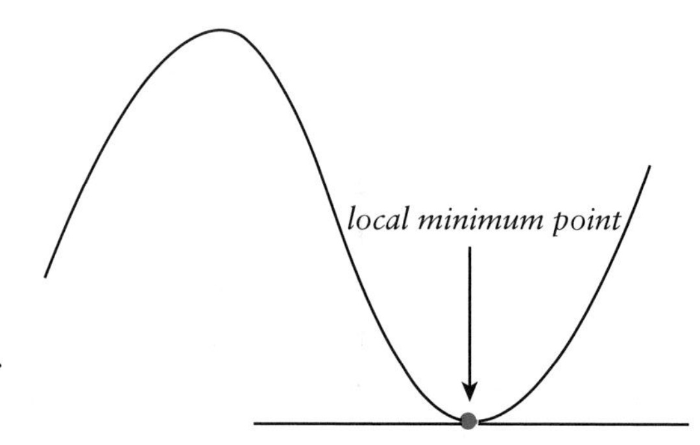 1. $\frac{dy}{dx} = 0$. 2. $\frac{dy}{dx}$ is negative just to the left of the point and positive just to the right of it Or $\frac{d^2y}{dx^2} > 0$ at the local minimum. It is a turning point. It belongs to a category of points called stationary points, at which $\frac{dy}{dx} = 0$. See also local maximum point and point of inflection.
locus	The set of all points, (x, y), which satisfy a specific condition. This condition can be used to create an equation which is satisfied by all points on the locus. The locus of all points that are equidistant from a fixed point is a circle. The locus of all points that are equidistant from two fixed points is a line. The line is the perpendicular bisector of the line segment joining the two points. See also locus method.

locus method

A method for finding the equation of a locus.

We let (x, y) represent any point on the locus and then use the given condition to set up an equation that is satisfied exclusively by the points on the locus.

The formula $y - y_1 = m(x - x_1)$ for the equation of a line and the formula $(x - h)^2 + (y - k)^2 = r^2$, for the equation of a circle, are established using the locus method.

log

Abbreviation for logarithm. $\log_a b$ is the index that must be placed on a to give b.
For example, $\log_2 8 = 3$ since $2^3 = 8$.

See also natural log (logarithm) and common log.

logarithm

See log.

loss

The difference between the selling price and the cost price when the cost price is greater.

See also profit.

lower quartile

A figure obtained from a cumulative frequency curve, as follows:

- Divide the highest frequency by 4
- Find this number on the vertical axis and draw a horizontal line to hit the curve
- Now draw a vertical line to hit the horizontal axis
- The number hit on the horizontal is the lower quartile

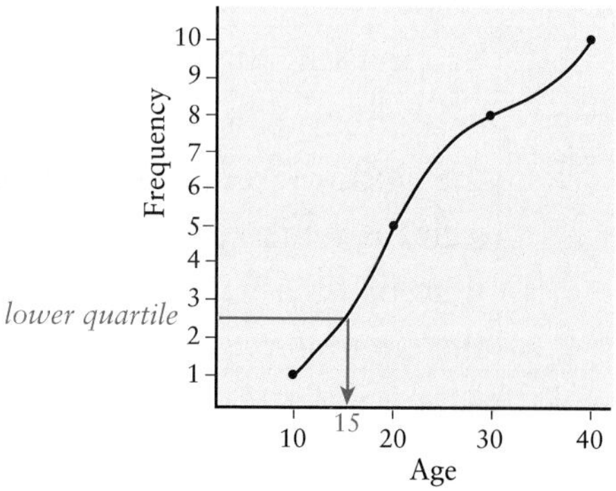

Frequency
10
9
8
7
6
5
4
3
2
1
lower quartile
10
15
20
30
40
Age

In this example, the lower quartile is 15.

See also upper quartile, median (statistics) and interquartile range.

lowest common multiple	The smallest number into which two or more numbers will divide. Abbreviated to lcm. For example, the lowest common multiple of 3 and 4 is 12. See also **multiple**, **factor** and **highest common factor**.

Mm

m	Abbreviation for **metre**.
Maclaurin	Colin Maclaurin (1698–1746). Scottish mathematician who went to Glasgow University at age eleven. By the age of nineteen, he was a Professor of Mathematics at Aberdeen. He is best known for the **Maclaurin series**, which appeared in his response to **Berkeley**'s criticism of the foundations of the theory of **calculus**.
Maclaurin series	An **infinite**, **convergent power series** in the form: $a_0 + a_1x + a_2x^2 + a_3x^3 + \ldots\ldots\ldots$, where $a_n = \frac{1}{n!}f^n(0)$ and $f^n(x)$ is the nth derivative of $f(x)$.
main diagonal	In a **matrix**, the **diagonal** from top left to bottom right.
major arc	A continuous part of the **circumference** of a **circle**, which is more than half the length of the circumference. See also **minor arc** and **semi-circle**.
map	See **transformation**.

matrix

An array of numbers designed to make certain calculations easier to perform.
An $m \times n$ matrix has m rows and n columns.

$\begin{pmatrix} 5 \\ 8 \end{pmatrix}$ is a 2×1 matrix.

The simultaneous equations $2x - 3y = 1$, $x + 5y = 3$

can be solved using matrices by rewriting the problem:

$$\begin{pmatrix} 2 & -3 \\ 1 & 5 \end{pmatrix}\begin{pmatrix} x \\ y \end{pmatrix} = \begin{pmatrix} 1 \\ 3 \end{pmatrix}$$

See also inverse of a matrix.

mean

See arithmetic mean.

See also geometric mean and weighted mean.

median (coordinate geometry)

In coordinate geometry, the median is a line from a vertex to the midpoint of the opposite side.

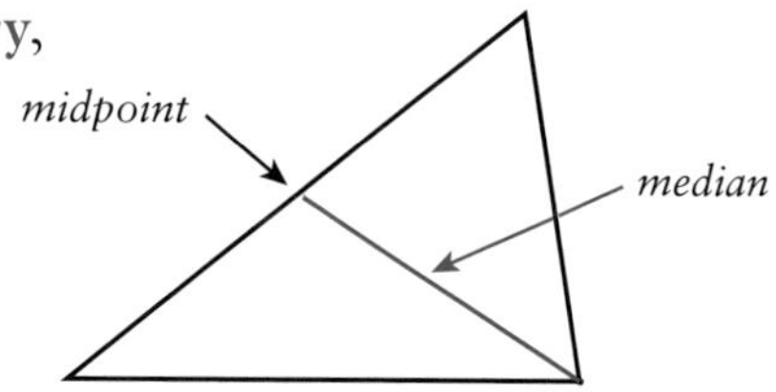

See also centroid and mediator of a line segment.

median (statistics)

The middle item in a list of numbers when they are arranged in order. The median may be estimated from a cumulative frequency curve, as follows:

- Divide the highest frequency by 2
- Find this number on the vertical axis and draw a horizontal line to the curve
- At this point, draw a vertical line to hit the horizontal
- The number hit on the horizontal is an approximation to the median

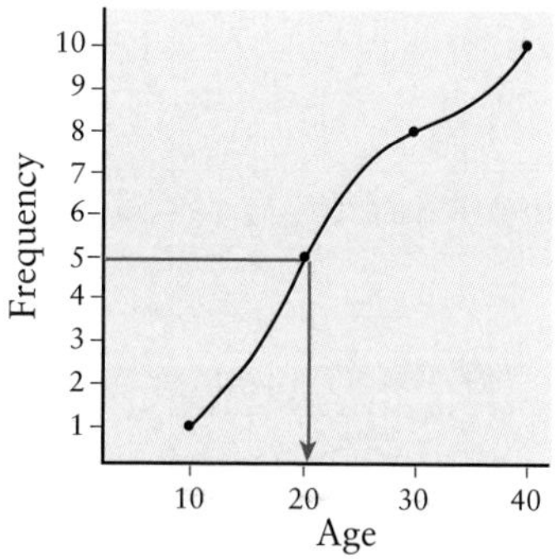

In this example, the median is approximately 20.

See also mean and mode.

mediator of a line segment	Also known as the perpendicular bisector of a line segment. It is a line, perpendicular to the line segment, which passes through the midpoint of the line segment. Every point on the mediator of a line segment [*ab*] is equidistant from *a* and *b*. 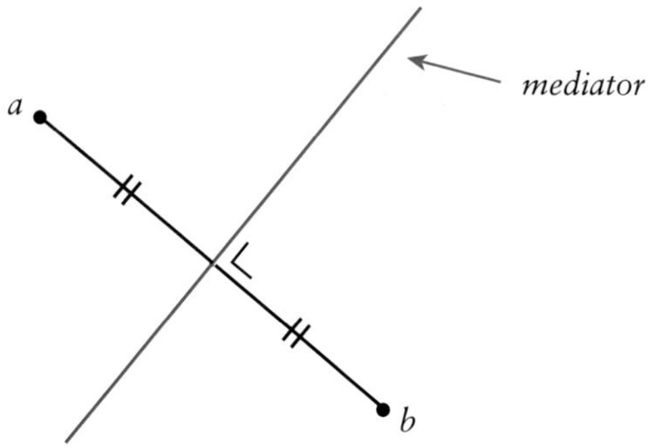 In a triangle, the mediators of the sides are concurrent and their point of intersection is called the circumcentre of the triangle. See also circumcircle.
member	See element.
metre	Unit of measurement. Abbreviated to m. 1 kilometre (km) = 1000 metres (m) 1m = 100 centimetres (cm) 1cm = 10 millimetres (mm)
mg	Abbreviation for milligram.
midpoint	The point on a line segment which is equidistant from the endpoints.
midpoint formula	In coordinate geometry, the formula $\left(\frac{x_1 + x_2}{2}, \frac{y_1 + y_2}{2}\right)$ is used to find the coordinates of the midpoint of the line segment joining (x_1, y_1) to (x_2, y_2). See also division in a ratio formula.

milligram	Unit of mass. Abbreviated to mg. 1 **kilogram** (kg) = 1000 **grams** (g) 1g = 1000 milligrams (mg)
millilitre	Unit of **volume**. Abbreviated to m*l*. 1 **litre** (*l*) – $1000cm^3$ 1*l* = 1000 millilitres (m*l*)
millimetre	Unit of measurement. Abbreviated to mm. 1 **kilometre** (km) = 1000 **metres** (m) 1m = 100 **centimetres** (cm) 1cm = 10 millimetres (mm)
minor arc	A continuous part of the **circumference** of a **circle**, which is less than half the length of the circumference. See also **major arc**.
m*l*	Abbreviation for **millilitre**.
mm	Abbreviation for **millimetre**.
mod	In **groups**, an abbreviation of **modulo**. 15 = 1 mod(7) means that 15 gives a remainder of 1 when divided by 7. In general, if $a = b$ mod(c), then $a - b$ is a multiple of c.
mode	In **statistics**, the item that occurs most frequently. For example, 2 is the mode of the following numbers: 1, 2, 4, 3, 2, 4, 2, 5. See also **mean** and **median**.
modulo	See **mod**.

modulus	The modulus of a number, also called **absolute value**, written $\|x\|$, is the distance from the number to 0. Hence, $\|-3\| = 3$ and $\|3\| = 3$. Note: If $\|x\| = 2$, then $x = 2$ or $x = -2$. The modulus of a **complex number** is the distance from the number to $0 + 0i$. It can be calculated by $\|a + bi\| = \sqrt{a^2 + b^2}$ The modulus or **norm** of a **vector** is the length of the vector: $\|a\vec{i} + b\vec{j}\| = \sqrt{a^2 + b^2}$.
de Moivre	Abraham de Moivre (1667–1754). A French-born mathematician, de Moivre emigrated to England in 1685 where he became a tutor. He was one of the pioneers of probability theory, but is more famous for proving the important result used in complex numbers:  $(\cos\theta + i\sin\theta)^n = \cos n\theta + i\sin n\theta$, where $i = \sqrt{-1}$.
de Moivre's theorem	A **theorem** used to raise a **complex number** in **polar form** to a **power**. It states $(\cos\theta + i\sin\theta)^n = \cos n\theta + i\sin n\theta$, where $i = \sqrt{-1}$.
multiple	A number, a, is a multiple of another number, b, if b divides evenly into a. Remember that every number is a multiple of itself. For example, multiples of 6 are 6, 12, 18, 24……… See also **lowest common multiple**, **factor** and **highest common factor**.

mutually exclusive events	In probability, two events are said to be mutually exclusive if they cannot both be the result of a single experiment. For example, if a card is chosen at random from a deck, it cannot be both a heart and a diamond.

Nn

Napierian log	See natural log (logarithm).
natural log (logarithm)	Log whose base is e. Also called Napierian log.
natural numbers	The positive whole numbers, including 0, i.e. 0, 1, 2, 3, 4......... The natural numbers are represented by the letter N. The symbol N_0 represents all the natural numbers except 0. See also integer, rational numbers, real numbers and complex numbers.
negative	A negative number is smaller than 0. Examples are -1, $-\frac{3}{4}$, -54, etc.
negative angle	An angle that is measured in a clockwise direction. When representing complex numbers in polar form we use a negative value for the argument when the imaginary part is negative. $-70°$
negative sense of the x-axis	The side of the x-axis (or real axis in complex numbers) to the left of any line that crosses it.

net income	The part of a salary left after all deductions (**income tax** etc.) have been made. It is also called 'take home pay.' If the only deduction is income tax, then Net Income = Gross Income – Net Tax.
net tax	The amount of tax that a person actually pays. It is calculated by subtracting **tax credits** from **gross tax**. Net Tax = Gross Tax – Tax Credits
Newton	Newton, Isaac (1642–1727). English physicist and mathematician who discovered the **binomial theorem**, differential calculus and integral **calculus**. Also famous for his great work on mechanics, the *Principia*.
Newton-Raphson formula	Used in **calculus** to approximate **roots** of an **equation** $f(x) = 0$: $x_{n+1} = x_n - \dfrac{f(x_n)}{f'(x_n)}$ See also **iterative procedure**.
norm	See **modulus**.
normal	In **geometry**, a **line**, **perpendicular** to a given line. *tangent* *normal*

normal curve A symmetrical bell-shaped curve representing the probability density function of a normal distribution.

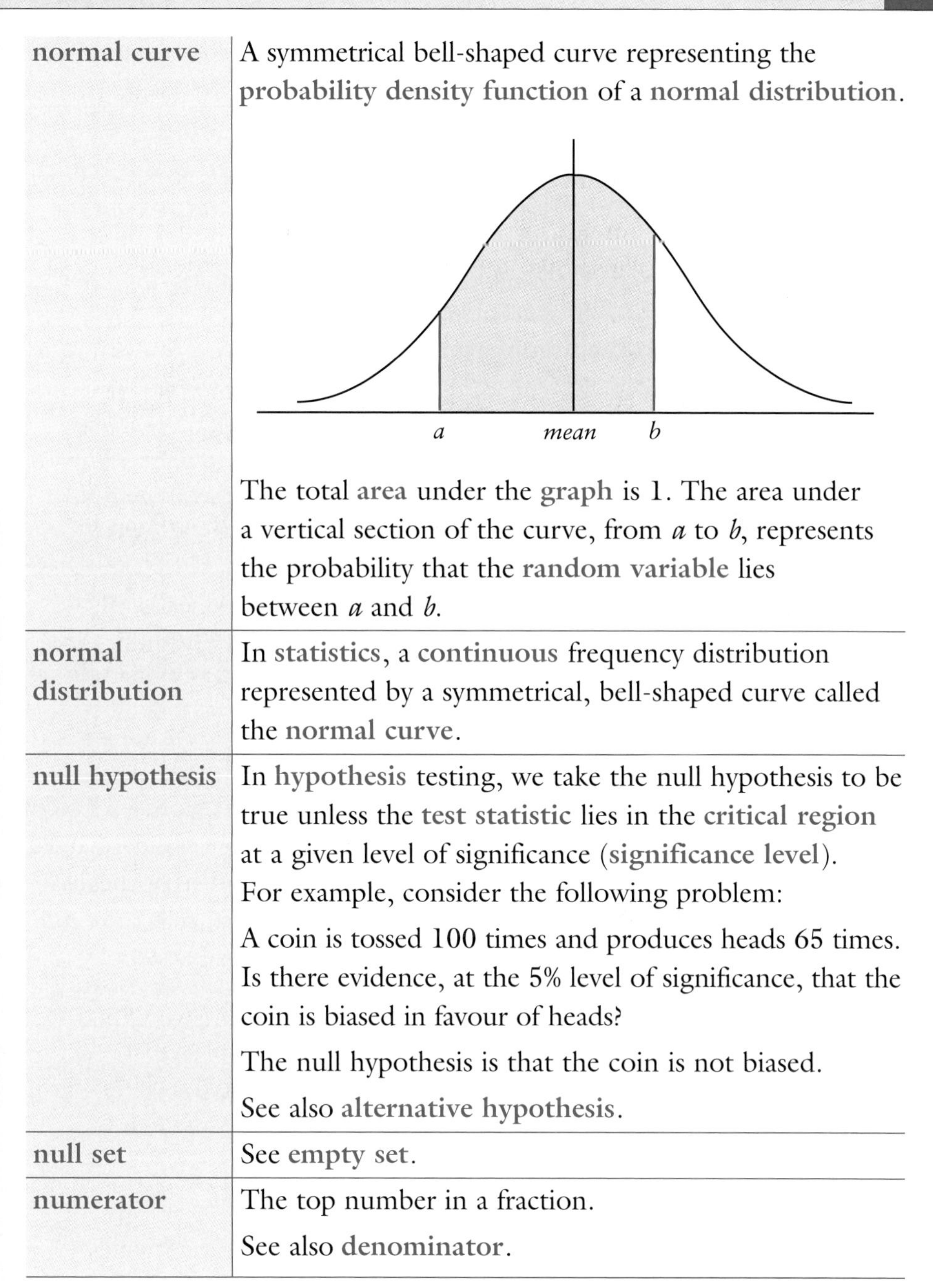

The total area under the graph is 1. The area under a vertical section of the curve, from a to b, represents the probability that the random variable lies between a and b.

normal distribution In statistics, a continuous frequency distribution represented by a symmetrical, bell-shaped curve called the normal curve.

null hypothesis In hypothesis testing, we take the null hypothesis to be true unless the test statistic lies in the critical region at a given level of significance (significance level). For example, consider the following problem:

A coin is tossed 100 times and produces heads 65 times. Is there evidence, at the 5% level of significance, that the coin is biased in favour of heads?

The null hypothesis is that the coin is not biased.

See also alternative hypothesis.

null set See empty set.

numerator The top number in a fraction.

See also denominator.

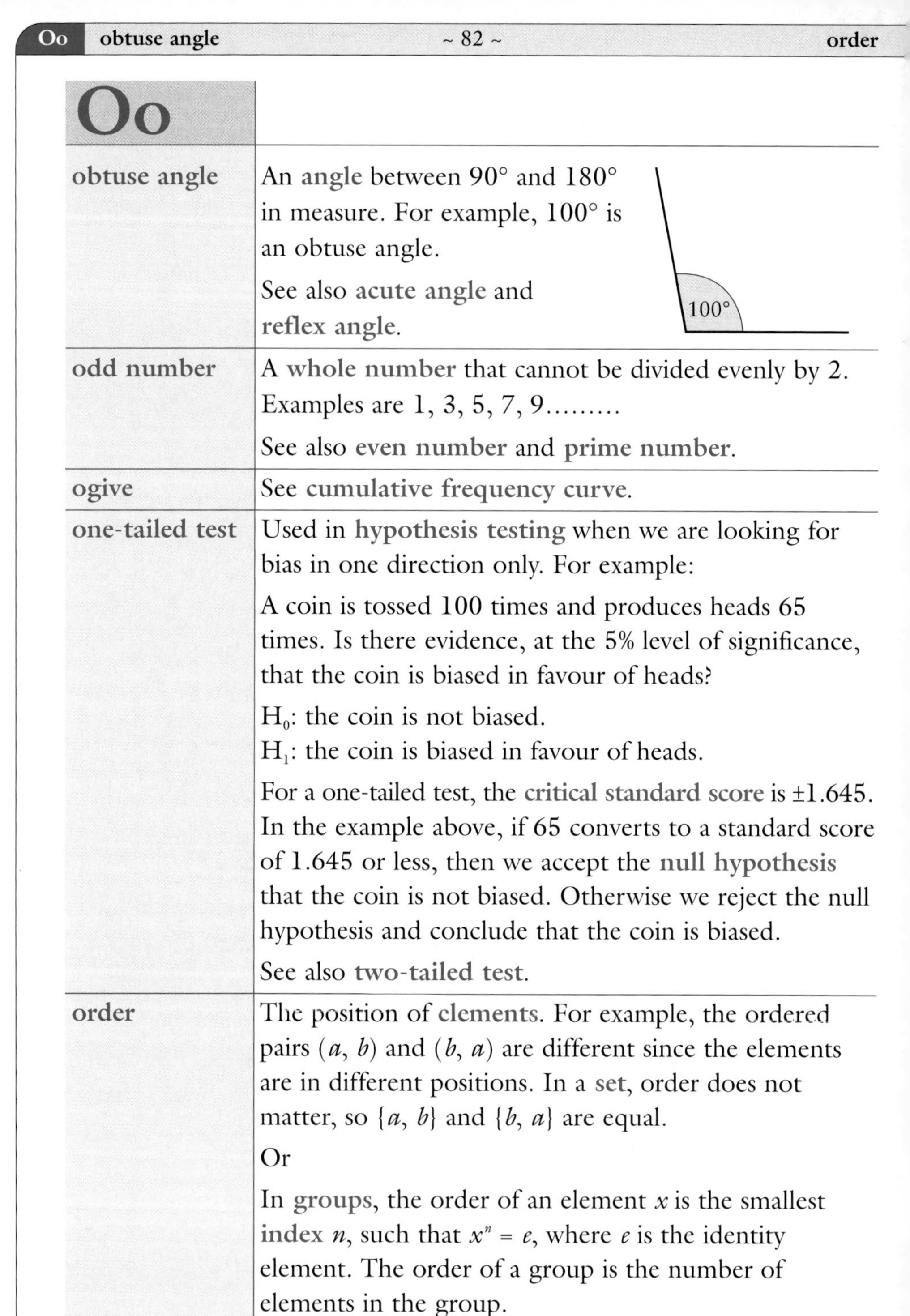

Oo

obtuse angle

An angle between 90° and 180° in measure. For example, 100° is an obtuse angle.

See also acute angle and reflex angle.

odd number

A whole number that cannot be divided evenly by 2. Examples are 1, 3, 5, 7, 9.........

See also even number and prime number.

ogive

See cumulative frequency curve.

one-tailed test

Used in hypothesis testing when we are looking for bias in one direction only. For example:

A coin is tossed 100 times and produces heads 65 times. Is there evidence, at the 5% level of significance, that the coin is biased in favour of heads?

H_0: the coin is not biased.
H_1: the coin is biased in favour of heads.

For a one-tailed test, the critical standard score is ±1.645. In the example above, if 65 converts to a standard score of 1.645 or less, then we accept the null hypothesis that the coin is not biased. Otherwise we reject the null hypothesis and conclude that the coin is biased.

See also two-tailed test.

order

The position of elements. For example, the ordered pairs (a, b) and (b, a) are different since the elements are in different positions. In a set, order does not matter, so $\{a, b\}$ and $\{b, a\}$ are equal.

Or

In groups, the order of an element x is the smallest index n, such that $x^n = e$, where e is the identity element. The order of a group is the number of elements in the group.

origin	In coordinate geometry, the point (0, 0). It is the reference point from which all other coordinates are measured. For example, the point (–2, 3) is 2 units to the left of the origin and 3 units above it. 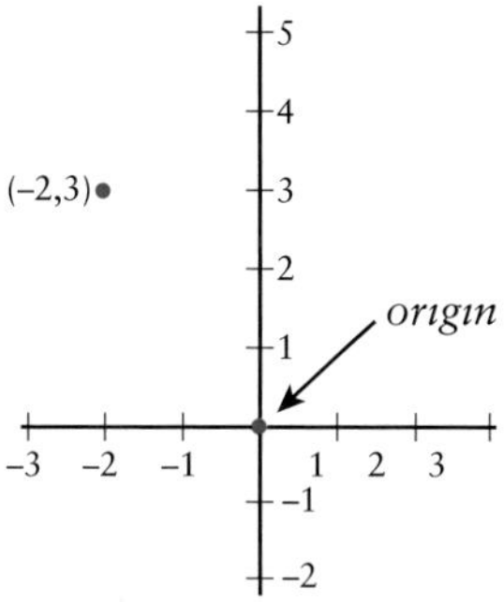
orthocentre	In a triangle, the point of intersection of the orthogonals. 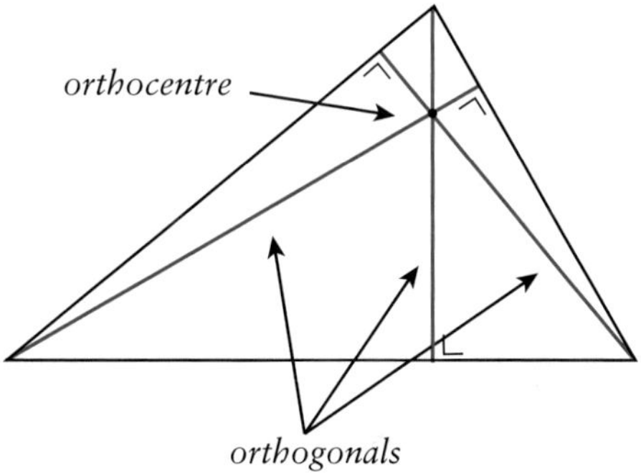 See also circumcentre, incentre of a triangle and centroid.
orthogonal	In a triangle, a line drawn from a vertex, perpendicular to the opposite side. 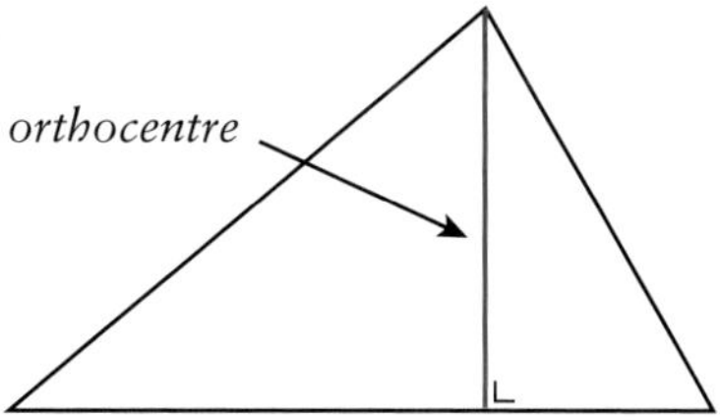 See also median, mediator of a line segment and bisector of an angle.

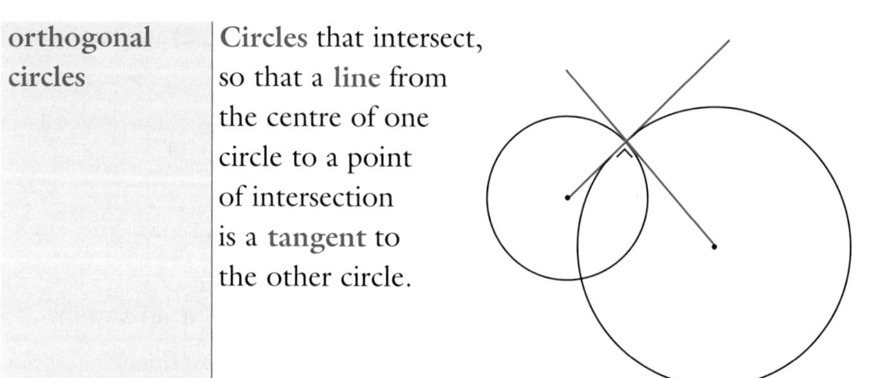

orthogonal circles

Circles that intersect, so that a line from the centre of one circle to a point of intersection is a tangent to the other circle.

Pp

parallel

Parallel lines are straight lines that are always the same distance apart.

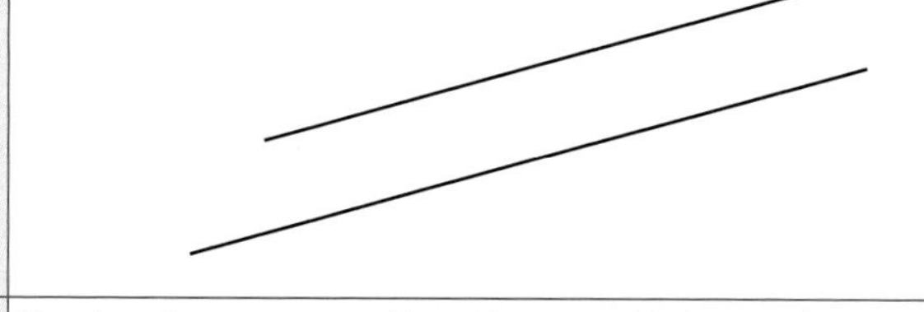

parallel projection

Projection onto a line L, parallel to a line K, is a set of couples (a, b) such that b is on the line L and the line ab is parallel to K.

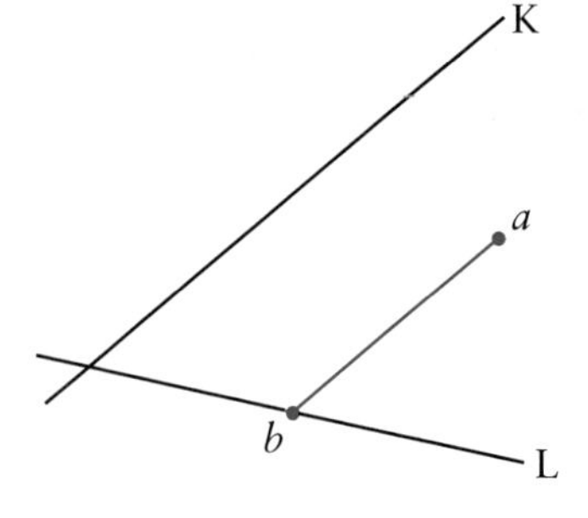

See also central symmetry, axial symmetry and translation.

parallelogram	A quadrilateral with opposite sides that are parallel. Opposite sides and opposite angles are always equal in measure. The diagonals always bisect each other. See also rhombus, rectangle, trapezium and square.
parallelogram law	A rule for adding two vectors by completing a parallelogram with the two vectors along adjacent sides and leaving the same point. The sum of the vectors is the diagonal from that point. $\vec{a}$ $\vec{a}+\vec{b}$ $\vec{b}$ See also triangle law.
parameter	A variable used in parametric equations. For example, in the equations $y = 3t - 1$ and $x = 2t$, t is the parameter.
parametric differentiation	A form of differentiation where we are asked to find $\frac{dy}{dx}$ but we are given x and y in terms of another variable, t, say. Use the formula: $\frac{dy}{dx} = \frac{dy}{dt} \div \frac{dx}{dt}$ *or* $\frac{dy}{dx} = \frac{dy}{dt} \times \frac{dt}{dx}$
parametric equations	Two equations that express two different variables, x and y, say, in terms of another variable, t, say. For example, in the equations $y = 3t - 1$ and $x = 2t$, t is called a parameter. See also Cartesian equation.

partial sum

The sum of an initial section of a **series**.
In the series u_1, u_2, u_3, u_4, u_5........., some partial sums are $u_1 + u_2$, and $u_1 + u_2 + u_3$, etc. The sum to infinity of the series exists if, and only if, the sequence of partial sums u_1, $u_1 + u_2$, and $u_1 + u_2 + u_3$ has a **limit**.

Pascal, Blaise

Blaise Pascal (1623–62).
French philosopher and mathematician who contributed to **calculus** and **probability**.
He invented and built one of the earliest calculating machines.

Pascal's theorem

$$\binom{n}{r} + \binom{n}{r+1} = \binom{n+1}{r+1}$$

This result is required to prove the **binomial theorem** and to build **Pascal's triangle**.

Pascal's triangle

The numbers in Pascal's triangle appear in the binomial expansions of $(x + y)^0$, $(x + y)^1$, $(x + y)^2$, $(x + y)^3$.........

Notice that each row begins and ends with 1. The other numbers in each row are obtained by adding the numbers to the left and right of it in the previous row.

1
1 1
1 2 1
1 3 3 1
1 4 6 4 1
1 5 10 10 5 1

See also **Pascal's theorem**.

pentagon

A five-sided figure. A regular pentagon has five sides equal in measure.
The angle between adjacent sides is 108°.

percentage

A **fraction** with 100 on the bottom, for example

$$75\% = \frac{75}{100} = 0.75$$

percentage error	When an approximation is made, Percentage Error = Relative Error × 100%. For example, if 13 is taken as an approximation to 12.8, then the **absolute error** is 0.2, the **true value** is 12.8 and **relative error** = $\frac{0.2}{12.8}$ = 0.015625. Percentage Error = 1.5625%.
perfect square	An **integer** obtained by squaring another integer. Examples are 1, 4, 9………
perimeter	The total **length** around the boundary of a two-dimensional figure. The perimeter of a **rectangle** = $2(length + breadth)$. The perimeter of a **circle** = $2\pi r$. See also **area** and **volume**.
period	The smallest **interval** after which a **periodic function** has a repeated set of values. For example, $\cos x$ is a periodic function with period 2π. See also **periodic graph**.
periodic function	A **function** that produces a set of values, repeated at regular intervals. It can be represented by a **periodic graph**. See also **period**.
periodic graph	A graph that repeats itself at regular intervals. It represents a **periodic function**. 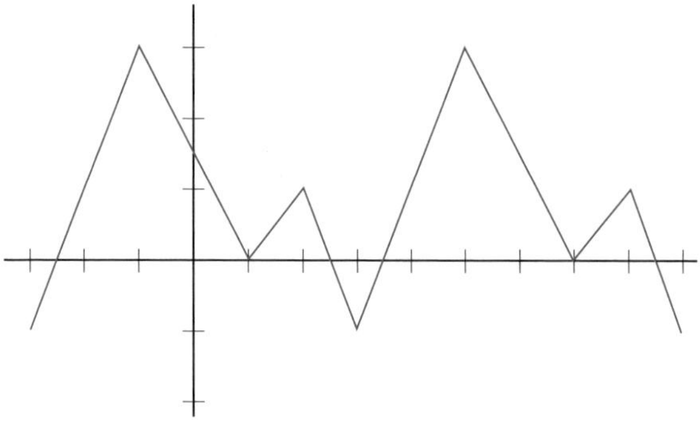 See also **period** and **range**.

permutation	An arrangement in a particular order. For example, abc and bac are two different permutations of the first three letters of the alphabet. The number of permutations of n different objects is given by the formula $n!$, if all objects must appear in each permutation. The number of permutations of r objects from n different objects is given by the formula $${}^{n}P_{r} = \frac{n!}{(n-r)!}.$$ See also combination.
perpendicular	Perpendicular lines form a right angle. In coordinate geometry, we prove that lines are perpendicular by multiplying the slopes. The answer should be –1. We prove that vectors are perpendicular by showing that their dot product is zero.
perpendicular bisector	See mediator.
perpendicular distance formula	In coordinate geometry, the formula $$\left\| \frac{ax_1 + by_1 + c}{\sqrt{a^2 + b^2}} \right\|$$ is used to find the perpendicular distance from the point (x_1, y_1) to the line $ax + by + c = 0$. See also distance formula.
perpendicular height	In a triangle, the perpendicular distance from a vertex to the opposite side. Perpendicular height is also known as the altitude of a triangle. *perpendicular height*

pi (π)	The ratio of the circumference of a circle to its diameter.
pie chart	In statistics, a circular chart in which the angle in each sector is proportional to the frequency.
plane	An infinite two-dimensional surface. Plane geometry is the study of objects that can be drawn on the plane, such as points, lines, angles, triangles, quadrilaterals and circles.
plane objects	Objects that can be drawn on the plane such as points, lines, angles, triangles, quadrilaterals and circles.
Plato	Plato (429–347 BC). Famous philosopher and mathematician. He founded Plato's Academy in Athens around 385 BC. Above the entrance to his academy lay the inscription 'Let no one ignorant of geometry enter here.' Plato's Academy brought together the best mathematical minds of the day. It flourished until 529 AD, when it was closed down by the Christian emperor Justinian, who claimed it was a pagan establishment. The most famous person associated with the academy was Aristotle. In his best-known work, *The Republic*, Plato outlines the mathematical syllabus that should be studied by the ruler of a state. It comprised arithmetic, plane geometry, solid geometry, astronomy and harmonics.

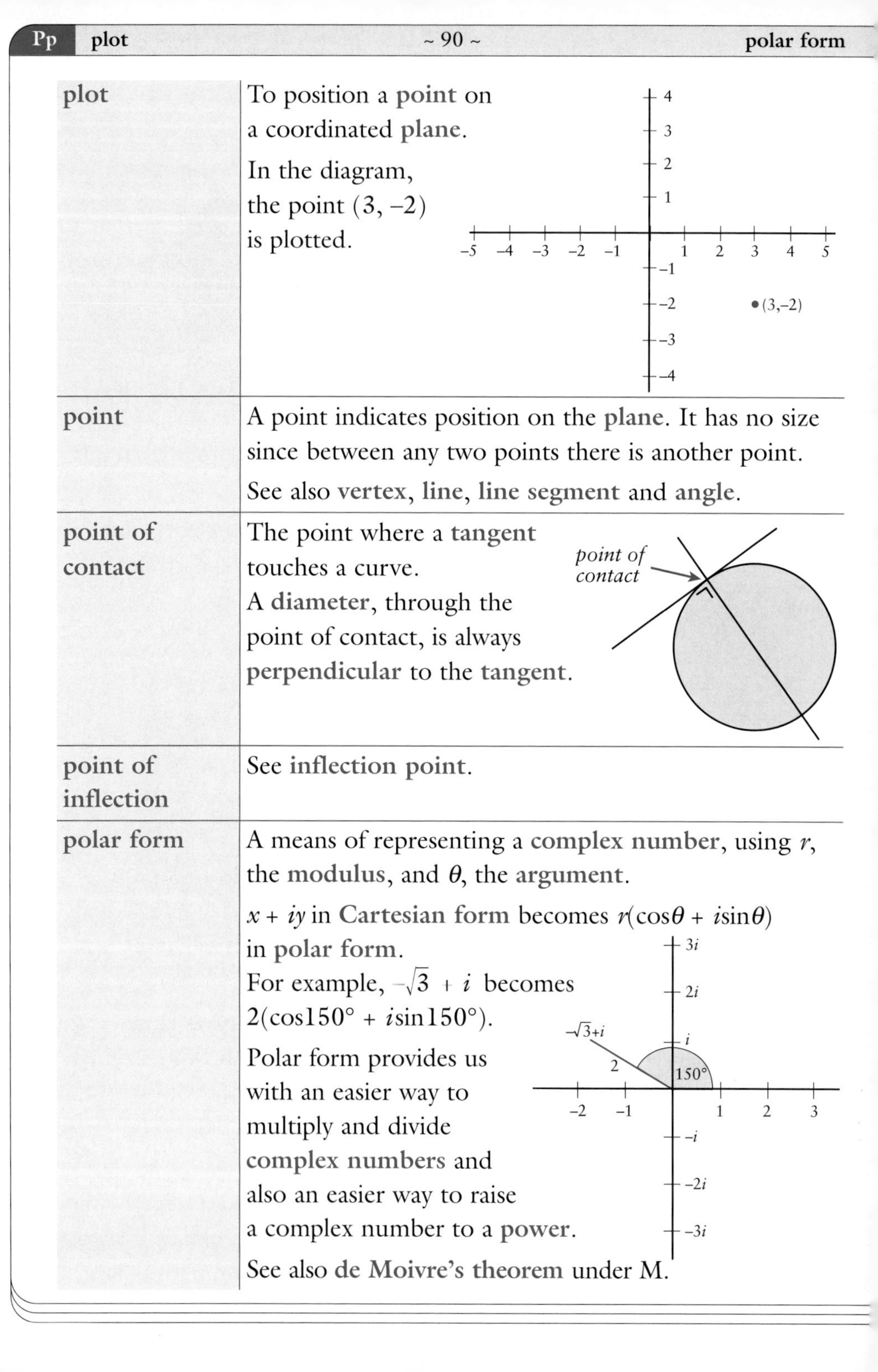

plot	To position a **point** on a coordinated **plane**. In the diagram, the point (3, –2) is plotted.
point	A point indicates position on the **plane**. It has no size since between any two points there is another point. See also **vertex**, **line**, **line segment** and **angle**.
point of contact	The point where a **tangent** touches a curve. A **diameter**, through the point of contact, is always **perpendicular** to the **tangent**.
point of inflection	See **inflection point**.
polar form	A means of representing a **complex number**, using r, the **modulus**, and θ, the **argument**. $x + iy$ in **Cartesian form** becomes $r(\cos\theta + i\sin\theta)$ in **polar form**. For example, $-\sqrt{3} + i$ becomes $2(\cos 150° + i\sin 150°)$. Polar form provides us with an easier way to multiply and divide **complex numbers** and also an easier way to raise a complex number to a **power**. See also **de Moivre's theorem** under M.

polygon	A plane figure with three or more sides. Examples of polygons are triangles, squares and hexagons.
polynomial	In algebra, an expression containing one or more variables. For example, $3x^4 - 2x^3 + x - 1$ is a polynomial in x.
population	In statistics, the set of figures from which samples are drawn.
positive	Greater than zero.
positive angle	An angle that is measured in an anti-clockwise direction. See also negative angle. 70°
positive sense of the x-axis	The side of the x-axis (or real axis in complex numbers) to the right of any line that crosses it.
post-multiplication	In matrices, to multiply on the right. For example, if the matrix A is post-multiplied by B, we get AB. See also pre-multiplication.
power	When a number is multiplied by itself, it is said to be raised to a power. For example, $2 \times 2 \times 2 \times 2$ is written 2^4 and is read '2 to the power of 4.' 2 is called the base and 4 is the index.
power series	A series in the form $a_0 + a_1x + a_2x^2 + a_3x^3 + ...$ See also Maclaurin series.
premise	Initial statement upon which a logical argument is based. See also theorem, deduction and conclusion.
pre-multiplication	In matrices, to multiply on the left. For example, if the matrix A is pre-multiplied by B we get BA. See also post-multiplication.

prime number

A whole number greater than 1 that has no factors except itself and 1. The first prime numbers are:
2, 3, 5, 7, 11, 13, 17, 19, 23, 29.........
1 is not a prime number.

See also **composite number**.

principal

A sum of money that has been invested, usually in a **compound interest** problem. The formula

$$A = P\left(1 + \frac{r}{100}\right)^n$$

tells us how much the principal, P, will amount to after n years at r% per annum.

See also **amount** and **interest**.

principal value

A single **element** of a **set** of numbers that obeys a particular rule.

For example, $\sin^{-1} 0$ is an angle whose sin is 0.
There is an **infinite** set of such angles:
{......–360°, 180°, 0°, 180°, 360°, ...}
The principal value of this set is 0°.

probability

A branch of mathematics dealing with the likelihood that an **event** will occur when an **experiment** is conducted. The probability of each outcome is expressed as a number between 0 and 1, with 0 representing an impossible outcome and 1 representing a certain outcome.

If an event has a probability of $\frac{1}{5}$, then that event should occur, on average, once out of every five times the experiment is conducted.

probability density function

A **function** representing the distribution of a **continuous random variable**. The integral of the function from a to b represents the **probability** that the random variable is between a and b.

product	The product of two or more numbers is obtained by multiplying them. Or In **algebra**, an **expression** where the last calculation is multiply. For example, $2x^2y$, $x(x + 1)$ and $(2x - 3)(x - 2)$ are products. When we **factorise** an expression, we convert it from a **sum** to a product. See also **sum**.
product rule	In **calculus**, the **formula** $$\frac{dy}{dx} = v.\frac{du}{dx} + u.\frac{dv}{dx}$$ is called the product rule. It is used to differentiate **products** where $y = uv$. See also **quotient rule** and **chain rule**.
profit	The difference between **selling price** and **cost price** where the selling price is greater. See also **loss**.
proof	Starting with a statement that is known to be true, a proof consists of a number of statements, each one following logically from the previous one, until the required **conclusion** is arrived at. See also **deduction** and **induction**.
proper fraction	A **fraction** in which the top number (**numerator**) is smaller than the bottom number (**denominator**). For example, $\frac{1}{2}$. See also **improper fraction** and **mixed number**.
proper subset	Any **subset** of a **set** apart from the **empty set** and the set itself. For example, $\{a, b\}$ is a proper subset of $\{a, b, c\}$. The symbol $\subset$ means 'is a subset of.' See also **improper subset**.

proportional	Two numbers x and y are proportional to two others a and b if $\frac{x}{y} = \frac{a}{b}$ See also **direct proportion**, **inverse proportion** and **ratio**.
Pythagoras	Pythagoras (c. 572–497 BC). Greek mathematician and philosopher, born on the island of Samos. In 540 BC Pythagoras founded a school of mathematics and philosophy at Crotona in southern Italy. Here, he gathered a group of disciples called the Pythagoreans. Bertrand Russell claimed that Pythagoras was 'intellectually, one of the most important men that ever lived.' The theorem famously attributed to Pythagoras (see **Pythagoras' theorem**) was, in fact, known to other cultures long before Pythagoras lived. It was this theorem that led to the discovery of **irrational** numbers. See also **Pythagoras' theorem** and **Pythagorean numbers**.
Pythagoras' theorem	In a **right-angled triangle**, the **square** on the **hypotenuse** is equal to the sum of the squares on the other two sides. $x^2+y^2=z^2$ See also **Pythagoras** and **Pythagorean numbers**.
Pythagorean numbers	Any three numbers that satisfy the **equation** $x^2 + y^2 = z^2$.

Qq

QED	Abbreviation for *quod erat demonstrandum*. Latin term meaning 'that which was to be proved.' It is often written at the end of a **proof** to indicate that the conclusion has been arrived at.
QEF	Abbreviation for *quod erat faciendum*. Latin term meaning 'that which was to be done.' It is often written at the end of a **construction** to indicate that the required construction has been completed.
quadrant	A **circle** is divided into four quadrants by **horizontal** and **vertical** lines through its **centre**. In **trigonometry**, it is important to know the quadrant in which an **angle** lies so that we can predict whether **sin**, **cos** and **tan** will be **positive** or **negative**. 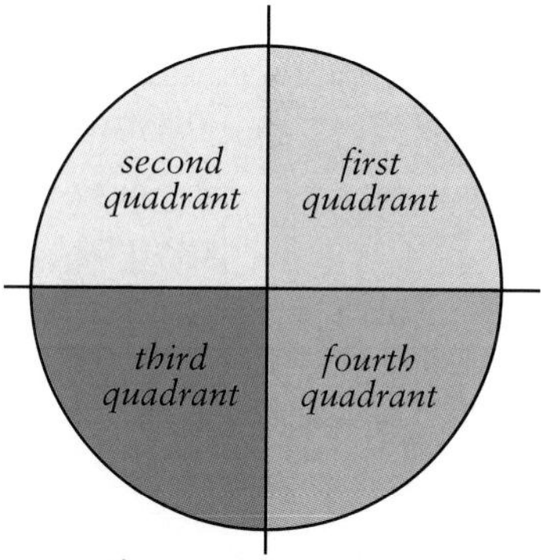 *second quadrant* · *first quadrant* · *third quadrant* · *fourth quadrant*
quadratic	Can apply to an **equation**, **expression** or **function** when the highest **power** of the **variable** involved is 2.
quadratic equation	In **algebra**, an equation that can be written in the form $ax^2 + bx + c = 0$. It is solved using **factors** or the **formula**: $x = \dfrac{-b \pm \sqrt{b^2 - 4ac}}{2a}$
quadratic equation formula	The **formula** $x = \dfrac{-b \pm \sqrt{b^2 - 4ac}}{2a}$ It is used to solve an **equation** in the form $ax^2 + bx + c = 0$. Note that if $b^2 - 4ac \geq 0$, the **roots** will be **real numbers**, whereas if $b^2 - 4ac < 0$, the roots will be **complex** numbers.

quadratic expression	In algebra, an expression in the form $ax^2 + bx + c$.
quadrilateral	Any four-sided plane figure. For example: See also cyclic quadrilateral, trapezium, parallelogram, rhombus, rectangle and square.
quartile	In statistics, quarter way divisions in the data when it is arranged in order. The difference between the lower quartile and the upper quartile is called the interquartile range. See also cumulative frequency curve.
quotient	A number obtained by division. For example, when 15 is divided by 2, the quotient is 7. See also remainder.
quotient rule	In calculus, the formula $\frac{dy}{dx} = \frac{v\frac{du}{dx} - u\frac{dv}{dx}}{v^2}$ where $y = \frac{u}{v}$ is used to differentiate fractions in the form $\frac{u}{v}$ and is called the quotient rule. If v is constant, then $\frac{dy}{dx} = \frac{\frac{du}{dx}}{v}$ See also product rule and chain rule.

Rr

radian

The angle at the centre of a circle standing on an arc, whose length is equal to the radius of the circle.

π radians = 180°

1 radian ≈ 57°

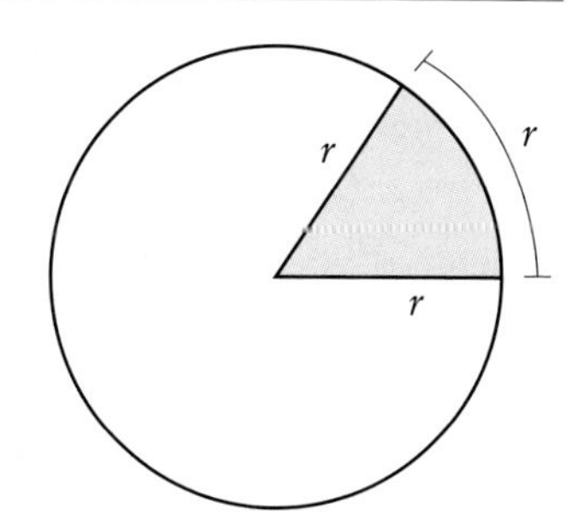

Radians provide an alternative to degrees for measuring angles.

radical axis

The line passing through the points of intersection of two circles.

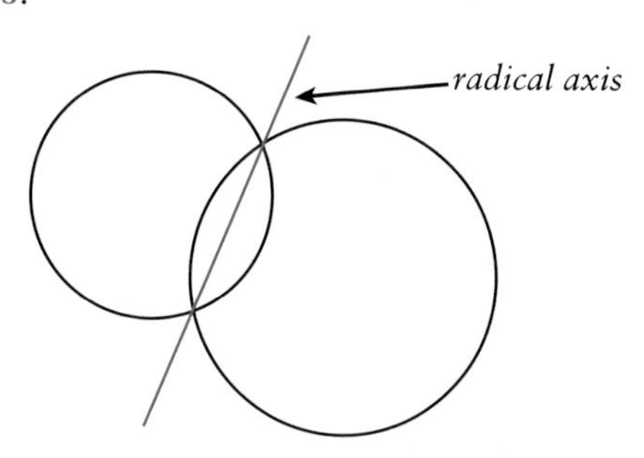

radius

The straight line joining the centre of a circle or sphere to any point on the circumference.

Or

The distance from the centre of a circle or sphere to any point on the circumference.

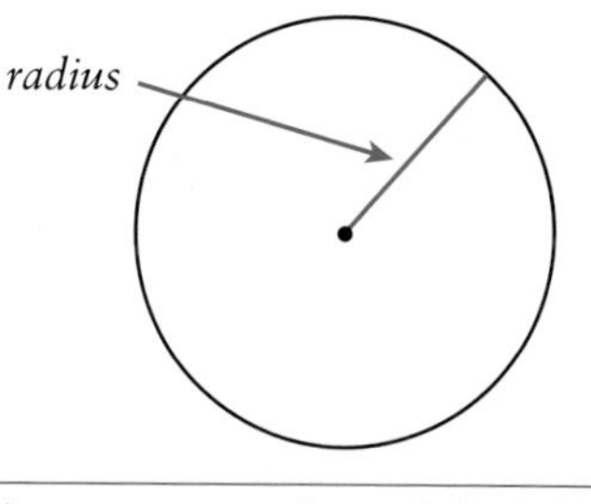

See also diameter.

random sample

In statistics, a sample that is representative of the population. Every element of the population should have an equal chance of being chosen in the sample.

See also sampling theory.

random variable	In **probability**, a random variable is a quantity that can have any of a certain range of **values**. The range of possible values will either be continuous or discrete. For example, if a die is thrown 500 times, the number of sixes that appear is a random variable. It can be any **whole number** between 0 and 500 and is therefore a **discrete random variable**. A random variable that can have any real value in a certain range is called a **continuous random variable**.
range	The **set** of second components in a **relation**. For example, in the relation {(1, 1), (2, 4), (3, 9), (4, 16)}, the range is {1, 4, 9, 16}. See also **domain** and **arrow diagram**.
rate of interest	The percentage interest. For example, if €1000 is invested at 4% **compound interest**, then the rate of interest is 4%. See also **principal** and **amount**.
rate of change	A **function** of time changes its value as time changes. The rate of change is the **ratio** of the change in value of the function to the change in time. For example, if **velocity** changes by 10 m/s over 2 seconds, then the rate of change of velocity is $\frac{10}{2} = 5\text{m/s}^2$. Velocity is the rate of change of distance and acceleration is the rate of change of velocity. A rate of change can be also found by **differentiation**. $\frac{dy}{dt}$ may be interpreted as the rate of change of y with respect to t.
rates	Taxes paid on property.

ratio	Measure of the relative size of quantities. A ratio is usually written by separating numbers with a colon(:), but can also be written as a **fraction**, which in turn can be written as a **decimal**. For example, if a:b = 2:1, then a is twice as big as b. A trigonometric ratio (sin, cos, tan, etc.) compares the lengths of sides in a **right-angled triangle**. See also **proportional**, **direct proportion** and **inverse proportion**.
ratio test	The test employed to determine if an **infinite series** is **convergent** or **divergent** when we cannot find an expression for S_n, the sum of the series. The ratio test states that if $\left\lvert \lim_{n \to \infty} \frac{u_{n+1}}{u_n} \right\rvert < 1$ then the series is convergent.
rational function	A **fraction** with a **polynomial** on top and/or bottom. For example, $\frac{1}{2x+3}$, $\frac{x^2 - x}{x+1}$ See also **function**.
rational numbers	Numbers that can be expressed in the form: $\frac{a}{b}, \quad a, b \in Z, b \neq 0$ Examples of rational numbers are $0, \frac{1}{2}, -\frac{4}{5}, 7\frac{1}{2}, 4.368, 100.$ The letter Q represents the set of rational numbers. See also **integer**, **natural numbers**, **real numbers** and **complex numbers**.
real axis	The **horizontal** axis in an **Argand diagram**. See also **imaginary axis**.
real number line	A means of representing real numbers on a diagram. –9 –8 –7 –6 –5 –4 –3 –2 –1 0 1 2 3 4 5 6 7 8 9

real numbers	A set comprising **rational numbers** and **irrational numbers** which is represented by the letter R. The real numbers include **whole numbers**, **fractions**, **decimals**, **surds** and numbers like π and e. The real numbers do not include **imaginary numbers**. See also **integer**, **natural numbers**, **rational numbers** and **complex numbers**.
reciprocal	The reciprocal of x is $\frac{1}{x}$. The reciprocal of $\frac{a}{b}$ is $\frac{b}{a}$. For example, the reciprocal of 3 is $\frac{1}{3}$ and the reciprocal of $\frac{2}{5}$ is $\frac{5}{2}$. The **product** of a number and its reciprocal is 1.
rectangle	A **parallelogram** whose **adjacent sides** are **perpendicular**. 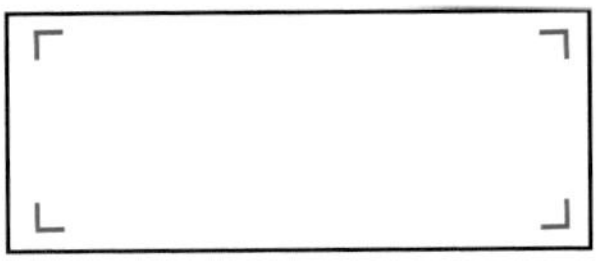 The **diagonals** of a rectangle are equal in length. This is not true of parallelograms that do not contain a right angle. See also **quadrilateral**, **trapezium**, **rhombus** and **square**.
recurrence relation	See **difference equation**.
recurring decimal	A **decimal** involving a certain number or numbers that repeat forever. For example, 23.17777777......... All recurring decimals are **rational numbers** and can, therefore, be represented as a fraction. The recurring part can be written as an **infinite** geometric series. For example, 0.7777777777 = 0.07 + 0.007 + 0.0007.........

recursive definition	A definition of a **sequence** that describes how to obtain a term using previous terms. For example, $u_{n+2} = 3u_{n+1} - u_n$. See also **difference equation**.
reference angle	The **angle** in the first **quadrant** used to solve **trigonometric equations**. For example, to solve the equation $\sin x = -\frac{1}{2}$, the reference angle is the **acute angle** whose **sin** is $\frac{1}{2}$, i.e. 30°. The other solutions between 0° and 360° can then be found by applying the following rules for the appropriate quadrants: 180° – reference angle | reference angle 180° + reference angle | 360° – reference angle In this case, since sin is negative in the third and fourth quadrants, the solutions are 210° and 330°.
reflection in a line	see **axial symmetry**.
reflection in a point	see **central symmetry**.

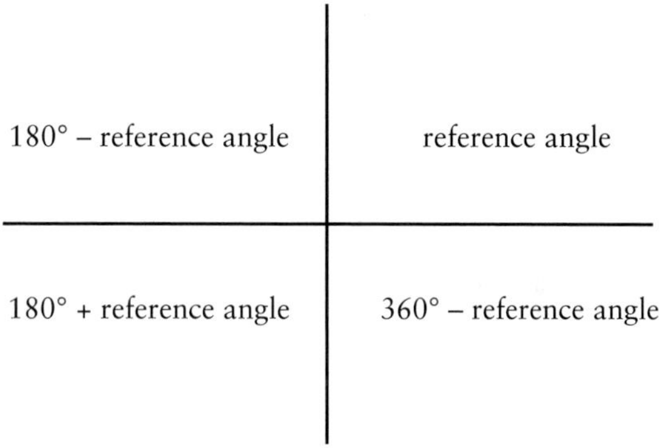

reflex angle	An angle bigger than 180° in measure. For example 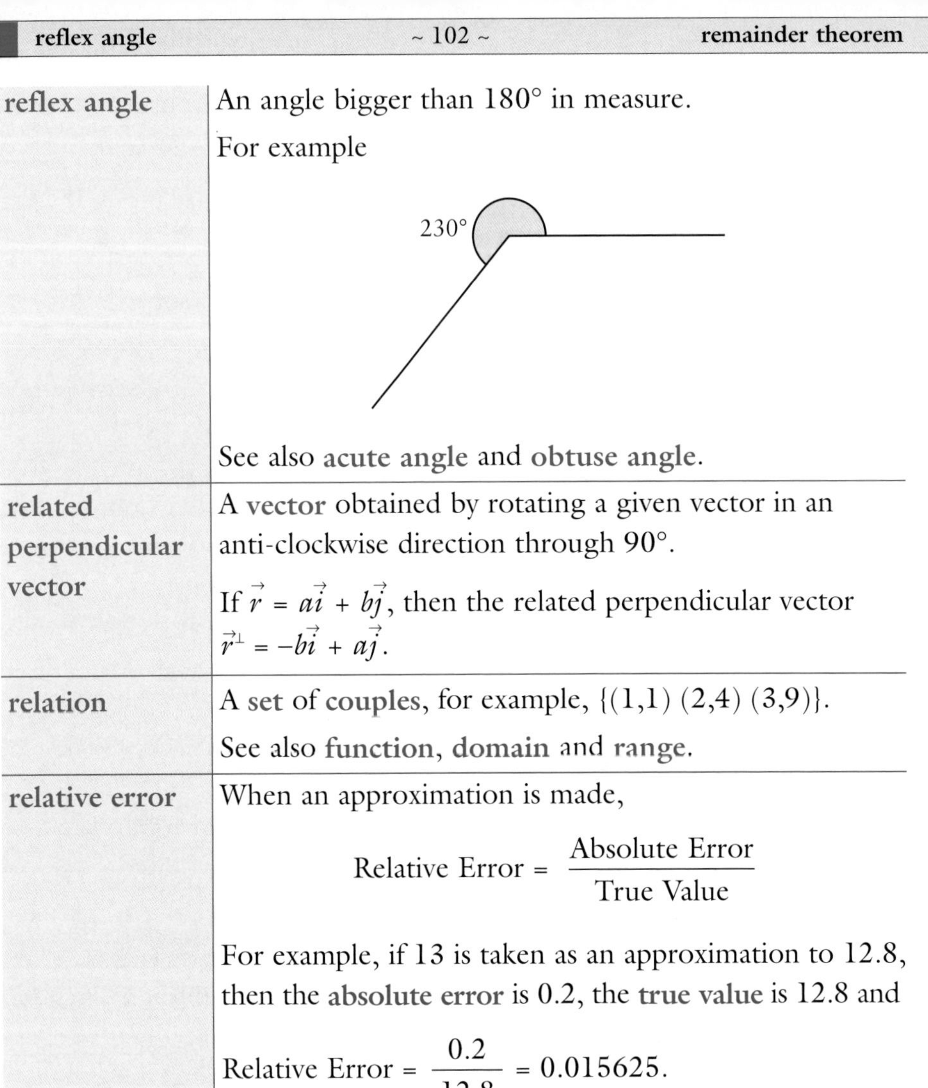 See also **acute angle** and **obtuse angle**.
related perpendicular vector	A **vector** obtained by rotating a given vector in an anti-clockwise direction through 90°. If $\vec{r} = a\vec{i} + b\vec{j}$, then the related perpendicular vector $\vec{r}^{\perp} = -b\vec{i} + a\vec{j}$.
relation	A **set** of **couples**, for example, {(1,1) (2,4) (3,9)}. See also **function**, **domain** and **range**.
relative error	When an approximation is made, $\text{Relative Error} = \dfrac{\text{Absolute Error}}{\text{True Value}}$ For example, if 13 is taken as an approximation to 12.8, then the **absolute error** is 0.2, the **true value** is 12.8 and $\text{Relative Error} = \dfrac{0.2}{12.8} = 0.015625.$
remainder	In division, the amount left over when one number does not divide evenly into another. For example, when 11 is divided by 3 the remainder is 2. See also **quotient**.
remainder theorem	When the **polynomial** $P(x)$ is divided by $x - a$, the **remainder** is $P(a)$. See also **factor theorem**.

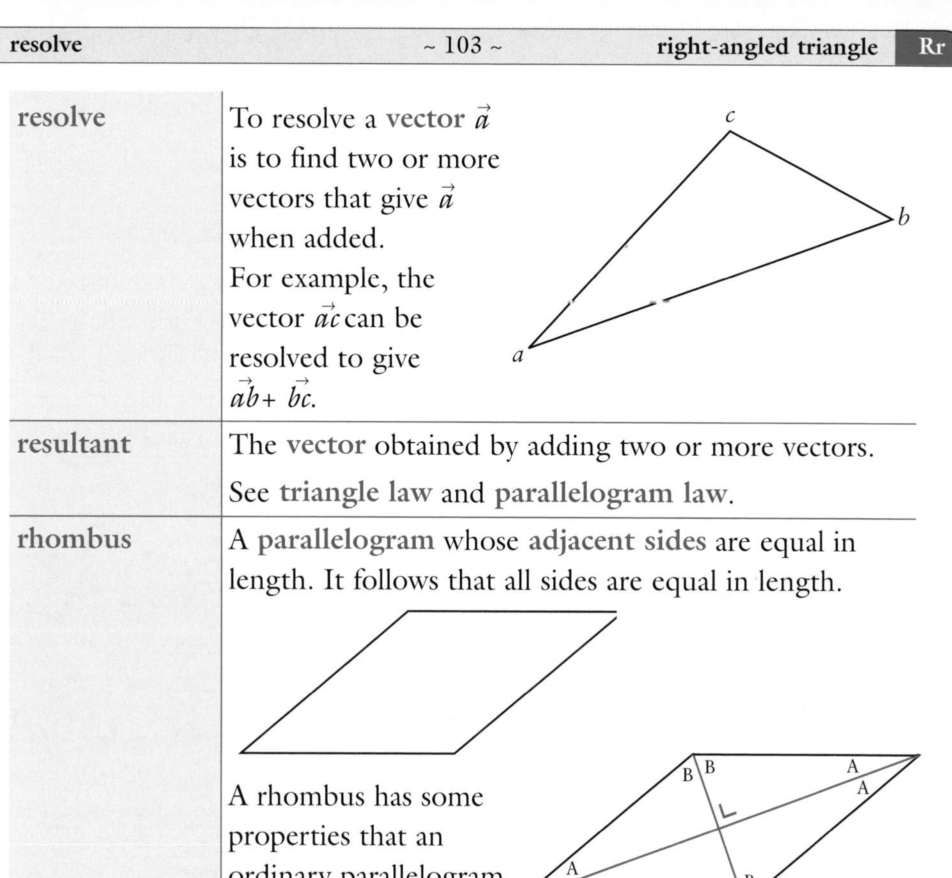

resolve To resolve a **vector** $\vec{a}$ is to find two or more vectors that give $\vec{a}$ when added.
For example, the vector $\vec{ac}$ can be resolved to give $\vec{ab} + \vec{bc}$.

resultant The **vector** obtained by adding two or more vectors.
See **triangle law** and **parallelogram law**.

rhombus A **parallelogram** whose **adjacent sides** are equal in length. It follows that all sides are equal in length.

A rhombus has some properties that an ordinary parallelogram does not have.
For instance, the **diagonals** are **perpendicular**.
Also, each **diagonal bisects** the angles in the corner of the rhombus.

right angle An angle of 90° in measure. The symbol ∟ in an angle represents a right angle. **Lines** that form a right angle are said to be **perpendicular**.

right-angled triangle A triangle, one of whose angles is 90° in measure. The **side** opposite the right angle is called the **hypotenuse**.

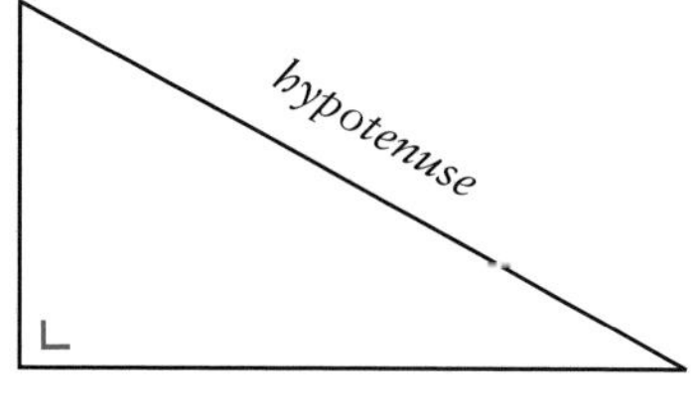

See also **Pythagoras' theorem**.

root	In algebra, the number which satisfies an equation, i.e. it makes the two sides equal when substituted for the variable. For example, 2 is a root of the equation $3x - 6 = 0$. The number of roots for a given equation is determined by the highest power of the variable, for instance a quadratic equation has two roots. See also extraneous roots.
rotation	An anti-clockwise rotation about a point p, through an angle θ, is a set of couples (a, b) such that 1. $\lvert\angle apb\rvert = \theta$ 2. $\lvert ab\rvert = \lvert bp\rvert$ 3. a, p, b are clockwise In the diagram, b is the image of a under a rotation about p through 70°. 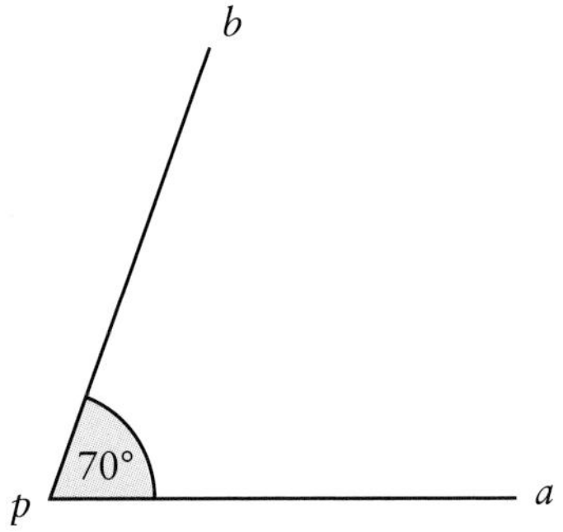 See also central symmetry, axial symmetry, translation and parallel projection.
rough estimate	The approximate result of a calculation, obtained by replacing each number involved with the nearest whole number. For example, $\frac{\sqrt{48} + 3.29}{1.6}$ could be estimated to be $\frac{7 + 3}{2} = 5$.
row	A horizontal array of numbers in a matrix.

Ss

saddle point	A **stationary point**, at which the **graph** crosses the horizontal **tangent**. Two conditions apply at a saddle point: 1. $\frac{dy}{dx} = 0$ since the tangent at a saddle point is horizontal. 2. $\frac{dy}{dx}$ has the same sign at a point on the graph to the left of the saddle point and at a point to the right of the saddle point. 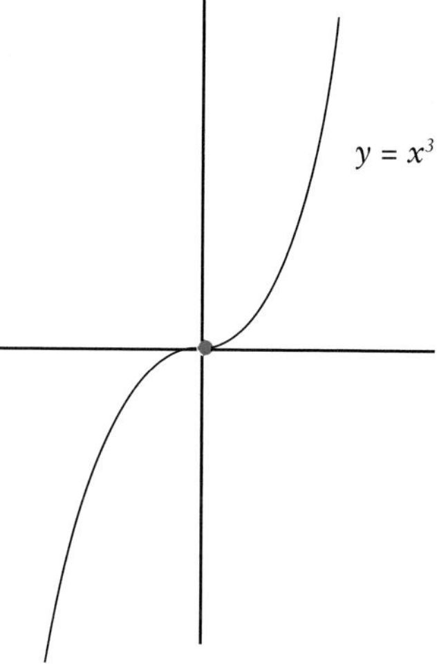 The **origin** is a saddle point on the **graph** of $y = x^3$. See also **local maximum** and **local minimum**.
sample	In **statistics**, a **set** of figures chosen from the **population**. The sample is usually random, meaning that every element of the population has an equal chance of being chosen for the sample. See also **sampling theory**.
sample space	In **probability**, the **set** of all possible outcomes when an **experiment** is conducted. For example, when a die is thrown the sample space is {1, 2, 3, 4, 5, 6}.
sampling theory	The branch of mathematics where a **sample** from a **population** is studied in order to make predictions concerning the entire population. The **mean** of the sample can be used to create a **confidence interval** for the true mean.

satisfy	A number is said to satisfy an **equation** if both sides are equal when the number is substituted for the **variable**. For example, 3 satisfies the equation $x^2 = 5x - 6$ since both sides are 9 when x is replaced by 3.
scalar	In **vectors**, we refer to quantities that have magnitude but not direction as scalars.
scalar product	See **dot product**.
scale factor	When an **enlargement** takes place, the scale factor is a number obtained by dividing the length of any side of the image object by the length of the corresponding side of the original object. 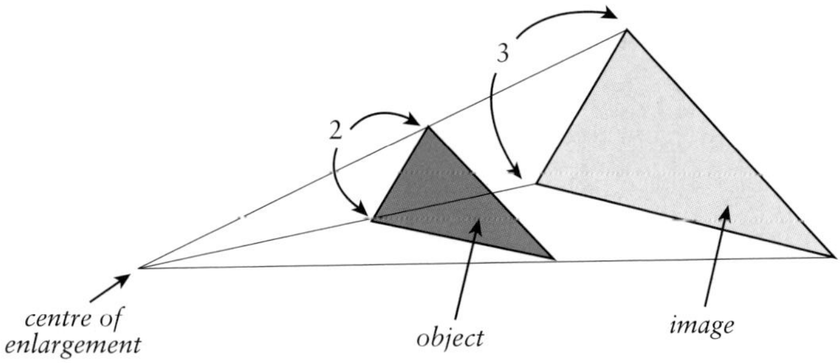 • In the above example, the scale factor is $\frac{3}{2} = 1.5$. Each side of the image is 1.5 times the length of the corresponding side of the object. • If the object is reduced in size then the scale factor is less than 1. • If k is the scale factor of an enlargement then area of image = area of object × k^2.
scalene triangle	A **triangle** with all **sides** of different lengths. The longest side is opposite the biggest **angle** and the shortest side is opposite the smallest angle. See also **isosceles triangle** and **equilateral triangle**.

scientific notation	Scientific notation (index notation) provides us with an alternative way to write numbers involving lots of zeros either at the beginning or at the end. For example, 0.0000002 or 256 000. Such numbers appear regularly in science. A number is said to be in scientific notation when it is in the form $a \times 10^n$, $1 \le a < 10$, $n \in Z$. The number 0.0000002 becomes 2.0×10^{-7} and 256 000 becomes 2.56×10^5 when written in scientific notation.
sec	In **trigonometry**, short for **secant**. See also **sin**, **cos**, **tan**, **cosec** and **cot**.
secant	In **trigonometry**, the ratio hypotenuse:adjacent in a **right-angled triangle**. Note: $\sec A = \dfrac{1}{\cos A}$ Or In **geometry**, a secant is a **line** that cuts a curve twice. *secant* See also **tangent**.
second derivative	Obtained by **differentiating** a **function** twice. It is represented by $\dfrac{d^2y}{dx^2}$ or $f''(x)$

second derivative test

In differentiation, a test used to determine if a **stationary point** is a **local maximum** or a **local minimum**. The x **coordinate** of the stationary point is substituted into the second derivative, $f''(x)$ or $\frac{d^2y}{dx^2}$, and the sign of the result tells us whether the point is a local maximum or a local minimum.

If $\frac{d^2y}{dx^2} < 0 \Rightarrow$ the point is a local maximum.

If $\frac{d^2y}{dx^2} > 0 \Rightarrow$ the point is a local minimum.

The test is inconclusive if $\frac{d^2y}{dx^2} = 0$.

In this case, the **first derivative test** should be used.

sector

Part of a **circle** enclosed by two radii and the **arc** that joins them.

The **area** of a sector is given by the **formula** $\frac{1}{2}r^2\theta$, where r is the **radius** of the circle and θ is the measure (in **radians**) of the angle in the sector.

See also **segment**.

segment

Part of a **circle** bounded by a **chord** and an **arc**.

See also **sector**.

selling price

The price of an article after **profit** is added or loss is subtracted.

Cost Price + Profit = Selling Price.

Cost Price – Loss = Selling Price.

See also **cost price** and **loss**.

semi-circle	Half of a circle. See also **sector**, **arc** and **segment**.
sequence	A list of numbers that are separated by commas and defined according to a definite rule. The numbers in the sequence are called **terms**. For example, 1, 3, 5, 7, 9……… is a sequence. u_n is usually used to represent an individual term in the sequence, where n represents the position of that term within the sequence. See also **series**.
series	A list of numbers added together, where the numbers are obtained according to a definite rule. The numbers in the series are called **terms**. For example, $1 + 3 + 5 + 7 + 9 + \dots\dots\dots$ is a series. S_n is used to represent the sum of the first n terms. See also **sequence**, **infinite series** and **finite series**.
set	A collection of clearly defined objects. The objects in the set are called **elements**. When listing the elements of a set, use chain brackets and separate the elements with commas, for example, {2, 3, 5, 9}.
set difference	A\B, the difference between the set A and the set B, is the set of elements that are in A but not in B. B\A is the set of elements that are in B but not in A. A B The shaded section in the **Venn diagram** represents A\B. For example, if A = {1, 3, 5, 7} and B = {1, 2, 3, 4} then A\B = {5, 7} and B\A = {2, 4}.

side	A line segment in a polygon. For instance, a square has four sides.
significance level	The probability of an error when hypothesis testing. For example, at the 5% level of significance the probability that we draw the wrong conclusion is 0.05.
significant figures	In any number the first non-zero digit on the left is the first significant figure. All other figures to the right of this are significant. For example: 23.003 has 5 significant figures. 0.00304 has 3 significant figures. See also decimal place.
similar triangles	See equiangular triangles.
simple interest	Interest that is calculated on the original principal and is the same each year. See also compound interest.
simple interest formula	$I = \frac{P \times R \times T}{100}$ where I is the simple interest, P is the sum of money invested, R is the rate of interest per annum and T is the number of years for which the sum is invested. See also compound interest formula.
simplify	In algebra, to rewrite an expression in a simpler way, thus changing the appearance of the expression but not its value. For example, $x^2 - x(2x - 1) - 3x + 2$ simplifies to $2 - 2x - x^2$. Simplify usually means 'remove brackets and add like terms.'

Simpson's rule	Used to **estimate** the **area** of an irregularly-shaped figure. The figure must be divided into an even number of vertical strips of width (w). The height of each strip is labelled h_1, h_2, h_3.......... Area = $\frac{w}{3}$ [first height + last height + 4(sum of even) + 2(sum of odd)] where even refers to heights h_2, h_4......... and odd refers to h_3, h_5..........
simultaneous equations	In **algebra**, two or more **equations** containing two or more **variables**. For example: $2x - y = 3$ $x + 3y = 4$ We solve the equations by gradually eliminating variables until we get an equation containing only one variable, from which its value can be found. This value is then substituted into another equation to find the value of a second variable and so on.
sin	In **trigonometry**, short for sine. In a **right-angled triangle**, sin is the **ratio** opposite:hypotenuse. See also **cos**, **tan**, **sec**, **cosec** and **cot**.
sine rule	In any **triangle**: $\frac{a}{\sin A} = \frac{b}{\sin B} = \frac{c}{\sin C}$ where a, b and c are the sides. A, B and C are the angles opposite those sides. The sine rule is used to find a missing **side** or **angle** in a triangle when we know a side, its opposite angle and another side or angle. See also **cosine rule**.
singular matrix	A **matrix** that does not have an inverse. The matrix $\begin{pmatrix} a & b \\ c & d \end{pmatrix}$ is singular if $ad - bc = 0$.

slope	The slope of a line is the tan of the angle that the line makes with the positive sense of the x-axis. In coordinate geometry, the slope of a line is obtained using the formula $$m = \frac{y_2 - y_1}{x_2 - x_1}$$ where (x_1, y_1) and (x_2, y_2) are two points on the line. If two lines are parallel, then they have the same slope, whereas if two lines are perpendicular, then the product of their slopes is –1. Lines that slope up from left to right have a positive slope. Lines that slope down have a negative slope. Horizontal lines have zero slope. The slope of a vertical line is undefined. Slope is also referred to as gradient.
solid of revolution	A three-dimensional figure formed by rotating a section of a curve around an axis. The formula $\pi\int_a^b [f(x)]^2 dx$ gives the volume of the solid formed by rotating the curve $y = f(x)$ about the x-axis between the lines $x = a$ and $x = b$.
solution	The values that satisfy an equation. Also known as roots.
solve	To find the solution to an equation.
sphere	A perfectly round three-dimensional object, with all points on its surface an equal distance from its centre. This distance is called the radius of the sphere. The volume of a sphere is given by the formula $$V = \frac{4}{3}\pi r^3.$$ The surface area is given by $A = 4\pi r^2$. See also hemisphere, cylinder and cone.

square	In geometry, a rectangle whose sides are equal in length. See also quadrilateral, parallelogram, trapezium and rhombus. Or To square a number is to multiply it by itself.
square matrix	A matrix with the same number of rows and columns.
square root	The square root of x, where $x > 0$ is a positive number which gives x when multiplied by itself. The symbol for square root is $\sqrt{\ }$. For example, $\sqrt{16} = 4$. The square root of a negative number gives rise to imaginary numbers involving i where $i = \sqrt{-1}$. See also cube root.
standard deviation	In statistics, a measure of the dispersion about the mean. The formula for standard deviation is $$\sigma = \frac{\Sigma f(x - \bar{x})^2}{\Sigma f}$$ where $\bar{x}$ is the mean, x is a statistic, f is the frequency and σ is the standard deviation.
standard error	The standard deviation of a set of sample means. The standard error is represented by the symbol $\sigma_{\bar{x}}$ and is given by the formula $$\sigma_{\bar{x}} = \frac{\sigma}{\sqrt{n}}$$ where σ is the standard deviation of the population and n is the size of the sample. See also sampling theory.
standard normal distribution	A normal distribution with mean 0 and standard deviation 1.

standard rate cut-off point	The sum of money below which a person pays the **standard rate of tax** and above which he/she pays the higher rate of tax. For example, a person has a **gross income** of €60,000. The standard rate cut-off point is €25,000. The standard rate of tax is 20% and the higher rate is 40%. The person's **gross tax** is calculated as follows: €25,000 @ 20% = €5,000 €35,000 @ 40% = €14,000 Gross tax = €19,000
standard rate of tax	The basic rate of tax that applies to the first part of a person's income, up to a certain point, called the **standard rate cut-off point**.
standard score	A result expressed in units of **standard deviation** from the **mean**. Usually represented by the letter z. The **formula** $z = \frac{x - \mu}{\sigma}$ converts a value of x to a standard score, where μ is the mean and σ is the standard deviation. For example, if $x = 70$, $\mu = 50$ and $\sigma = 10$, then $z = \frac{70 - 50}{10} = 2$ indicating that 70 is 2 standard deviations above the mean. A negative value for z indicates that the score is below the mean.
stationary point	A **point** on a curve where $\frac{dy}{dx} = 0$. The **tangent** to the curve is **horizontal** at a **stationary point**. There are three types: **local minimum**, **local maximum** and **saddle point**. See also **inflection point**.
statistics	The branch of mathematics concerned with the collection and interpretation of **data**.

straight angle	An angle of 180° in measure. 180° See also **right angle**.
subgroup	A **subset** of a **group** that is itself a group under the same **binary operation**.
subject of a formula	The letter that appears on its own in a **formula**. For example, V is the subject of the formula $V = \frac{1}{3}\pi r^2 h.$ Any formula can be rearranged to change the subject to a different letter.
subset	The **set** B is a subset of the set A if there is no **element** of B which is not also in A. The symbol ⊂ reads 'is a subset of.' For example, the set {2, 3} is a subset of the set {1, 2, 3, 4). Every set has two **improper subsets**: the empty set and the set itself. All other subsets are called **proper subsets**.
substitute	To replace a **variable** in an **equation** or **expression** with a number.
sum	To 'find the sum of' means to add. Or In **algebra**, a sum is an **expression** made up of two or more **terms** separated by a + or a –. For example, $x + 1$, $x^2y + z$, $2x(x + 1) - 3$ are sums. To convert a sum to a **product** is to **factorise**. In **trigonometry**, we have formulae that convert sums to products and products to sums. See also **product**.

sum of two cubes	In algebra, an expression in the form $x^3 + y^3$. Note: The factors are $(x + y)(x^2 - xy + y^2)$. See also difference of two cubes and difference of two squares.
sum to infinity	In an infinite series, to find the sum to infinity is to find the limit as n goes to infinity of S_n. This is written $\lim_{n \to \infty} S_n$ or S_∞. If this limit exists then the series is said to be convergent. Of particular importance is the fact that the sum to infinity of a geometric series is given by the formula $\frac{a}{1-r}$, where a is the first term and r is the common ratio, on condition that $-1 < r < 1$.
supplementary angles	Angles that add to 180°. See also complementary angles.
surd	An expression containing a square root: $\sqrt{\ }$.
surface area	The total outer area of all the surfaces of a three-dimensional object. See also area and volume.
symmetric group	A group made up of all the permutations of a given set.
symmetry	Exact likeness in shape about a given line (axial symmetry) or point (central symmetry).

Tt

tan	In trigonometry, short for tangent. In a right-angled triangle, tan is the ratio opposite:adjacent. See also sin, cos, sec, cosec and cot.

tangent	See tan. Or In geometry, a line that touches a curve at only one point, called the point of contact. 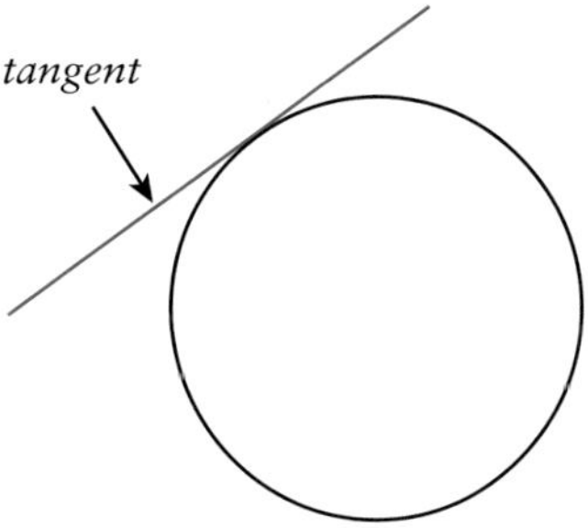 See also secant.
tax credits	Amount by which a person may reduce his/her gross tax bill. Net Tax = Gross Tax – Tax Credits See also net income and gross income.
telescoping series	A series in which each term may be written as the difference of two terms, allowing the sum of the series to be obtained through cancellation. For example, $\sum_{r=1}^{n} \frac{1}{r(r+1)} = \sum_{r=1}^{n}\left[\frac{1}{r} - \frac{1}{r+1}\right]$ $= \left[1-\frac{1}{2}\right]+\left[\frac{1}{2}-\frac{1}{3}\right]+\left[\frac{1}{3}-\frac{1}{4}\right]+ \ldots\ldots\ldots\ldots +$ $\left[\frac{1}{n-1}-\frac{1}{n}\right]+\left[\frac{1}{n}-\frac{1}{n-1}\right]$ $= 1 - \frac{1}{n+1}$ $= \frac{n}{n+1}$
term	In algebra, any combination of constants and/or variables that are multiplied together. The expression $2x^2y - x(2y - z)$ contains two terms. A bracket may be treated as an individual letter. Or An individual number in a sequence or series.

test statistic	In hypothesis testing, a statistic that has a known distribution under a null hypothesis, but a different distribution under an alternative hypothesis.
theorem	A mathematical proposition that can be deduced by logic from a set of axioms.
three-dimensional	Having length, width and depth.
top-heavy fraction	See improper fraction.
transcendental number	A real number that is not a root of any polynomial equation with rational coefficients. π and e are transcendental numbers.
transformation	A mapping or function, especially one which causes a change of shape or position in a geometric figure. Examples of transformations are central symmetry, axial symmetry, rotation, translation and linear transformation.
translation	A translation is a transformation that pushes a figure a given distance in a given direction without changing its orientation. If a point on the original figure is joined to its image, then the line segment will be the same length and pointing in the same direction as the translation. The diagram shows a triangle and its image under the translation $\overrightarrow{ab}$.

transpose of a matrix	A **matrix** formed by interchanging the rows and columns of a given matrix. For example, the transpose of $\begin{pmatrix} a & b \\ c & d \end{pmatrix}$ is $\begin{pmatrix} a & c \\ b & d \end{pmatrix}$
transversal	A **line** intersecting two or more lines.
trapezium	A **quadrilateral** containing two **parallel** sides of different lengths. 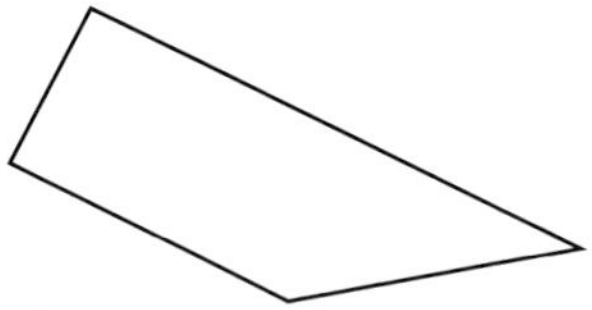 Note: The **area** of a trapezium is half the **sum** of the parallel sides multiplied by the **perpendicular** distance between them. See also **parallelogram**, **rectangle**, **rhombus** and **square**.
trend graph	In **statistics**, a **graph** that shows how a certain quantity changes over a certain period of time, measured at regular intervals. 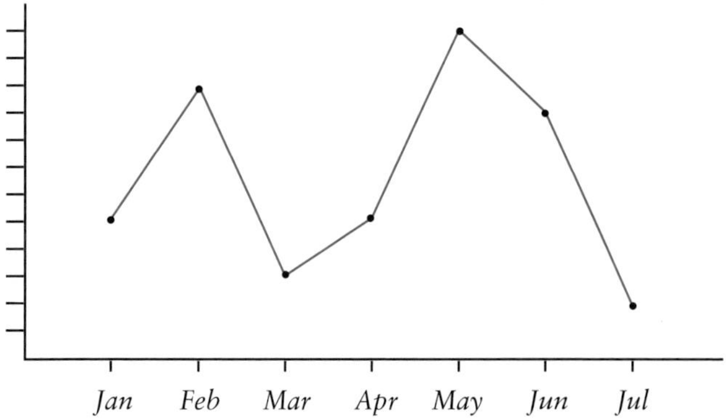 See also **bar chart**, **pie chart**, **histogram** and **cumulative frequency curve**.
trial	A single **experiment** in **probability**.

triangle

A three-sided plane figure.

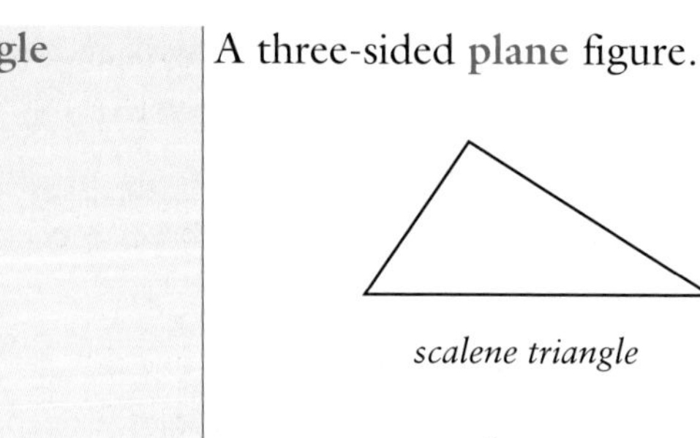
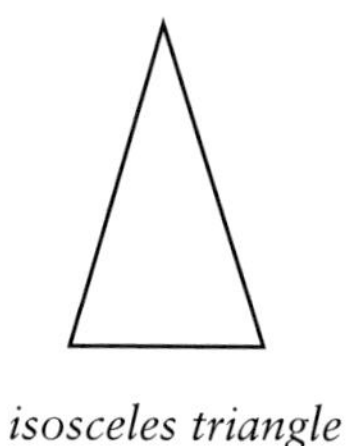

scalene triangle *isosceles triangle*

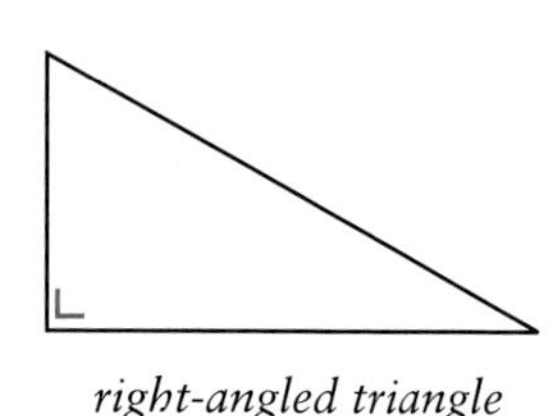

equilateral triangle *right-angled triangle*

The area of a triangle may be obtained using different formulae in different branches of mathematics.

In geometry: half the base multiplied by the perpendicular height.

In coordinate geometry: $\frac{1}{2}|x_1y_2 - x_2y_1|$

In trigonometry: half the product of any two sides multiplied by the sine of the angle between them, i.e.

$$\frac{1}{2}ab\sin C$$

triangle law

A method for adding vectors. One vector is positioned where the other ends and their sum is obtained by completing the triangle.

For example,

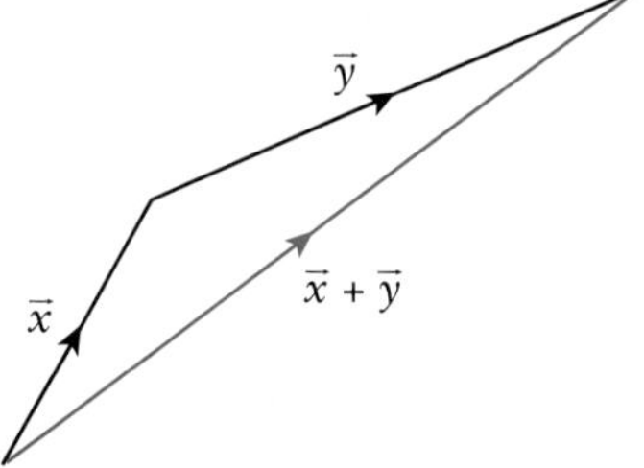

See also parallelogram law.

trigonometric ratios	Ratios that can be created from the lengths of sides in a right-angled triangle. The trigonometric ratios are sin, cos, tan, sec, cosec and cot.
trigonometry	The branch of mathematics dealing with problems related to triangles.
trinomial	An expression containing three terms, usually associated with factors. For example, $3x^2 - 2x + 5$.
trisect	To divide into three equal parts.
true value	When an estimate is made of a certain number, the true value is the exact value of the number.
turning point	A stationary point at which the graph does not cross the horizontal tangent. It can be either a local maximum point or a local minimum point. See also stationary point, saddle point and inflection point.
two-dimensional	Having length and width, but not depth.
two-tailed test	Used in hypothesis testing when looking for bias in either direction. For example: A coin is tossed 100 times and produces heads 65 times. Is there evidence, at the 5% level of significance, that the coin is biased? H_0: the coin is not biased. H_1: the coin is biased in favour of heads. For a two-tailed test, the critical region is < -1.96 and > 1.96 at the 5% significance level. In the above example, if 65 converts to a standard score between −1.96 and 1.96, then we accept the null hypothesis, that the coin is not biased. Otherwise we reject the null hypothesis and conclude that the coin is biased. See also one-tailed test.

Uu

unbiased	In **statistics**, a die is said to be unbiased if, when it is thrown, each number from 1 to 6 has an equal chance of appearing. See also **biased**.
union of sets	The **set** formed by joining the sets together without repeating an **element**. The symbol for union of sets is ∪. For example: If A = {1, 2, 3, 4} and B = {2, 4, 6, 8}, then A ∪ B = {1, 2, 3, 4, 6, 8}
unit	A physical quantity used as the basis of measurement. For example, a **centimetre** is a unit for measuring **length** and a **radian** is a unit for measuring **angles**.
unit circle	A circle of **radius** one **unit**, usually drawn with centre, (0, 0). 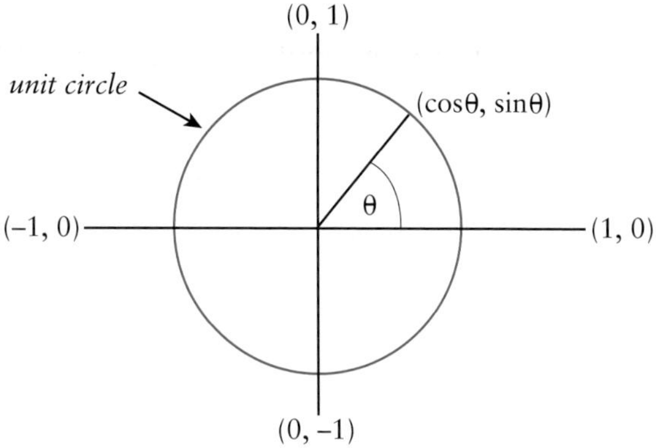
unit vector	A **vector**, one unit long. $\frac{\vec{x}}{\lvert\vec{x}\rvert}$ is the unit vector in the direction of $\vec{x}$.
unity	The number 1.

universal set	The set which contains all the required **elements** for a given problem. The universal set is sometimes called the universe and is represented by a rectangle on a **Venn diagram**.
universe	See **universal set**.
unlike terms	In **algebra**, **terms** that do not contain the same letters or do not have the same **index** on each letter. For example, x^2y and xy^2 are unlike terms. See also **like terms**.
upper quartile	A figure obtained from a **cumulative frequency curve**, as follows: • Divide the highest **frequency** by 4, then multiply by 3. • Find this number on the vertical axis and draw a horizontal line to hit the curve. • Now draw a vertical line to hit the horizontal axis. • The number hit on the horizontal is the upper quartile. In this example, the upper quartile is 27. 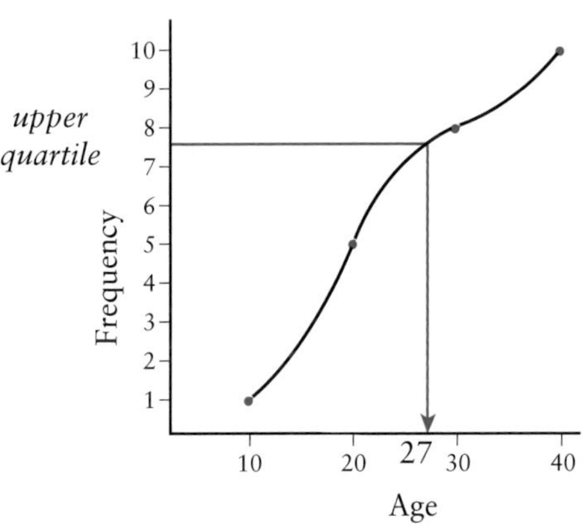 See also **lower quartile** and **interquartile range**.

Vv

variable	A quantity that can take on different values, as opposed to a **constant**, which is always the same. For example, in the expression $2x + 3$, x is a variable, 2 and 3 are constants.
VAT	Abbreviation for Value Added Tax. This tax applies to both goods and work and is calculated as a percentage of the initial cost and then added to the price to be paid. For example, when a car is serviced, VAT applies to the parts at one rate and to the labour at a different rate.
vector	A quantity that has both magnitude and direction. Non vectors are called **scalars**. A vector is represented by a **line segment** and an arrow. The length of the line segment reflects the magnitude of the vector and the arrow indicates direction. 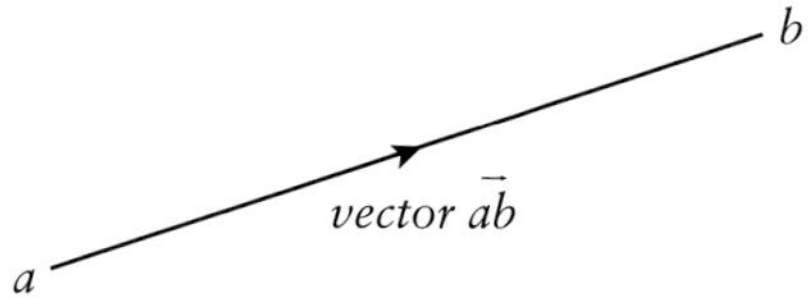
velocity	Velocity is speed in a given direction and is therefore a **vector** quantity. Speed is a **scalar** quantity since it has magnitude, but not direction.
Venn diagram	In **sets**, a **diagram** that uses **circles** or ovals to represent **sets** and a **rectangle** for the **universal set**. 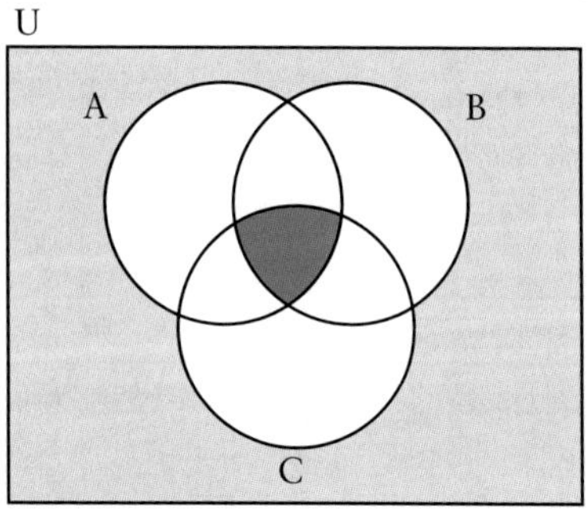
verify	Show to be true.

vertex	The point at which two sides of a polygon meet.
vertical	A line going straight up or down, forming a right-angle with the horizontal.
vertically opposite angles	Opposite angles when two lines intersect. They are equal in measure.
vertices	Plural of vertex.
volume	The amount of space taken up by a three-dimensional object. See also capacity.

Ww

weight	In statistics, a figure assigned to a number reflecting the significance of that number. Used to calculate the weighted mean.
weighted mean	In statistics, a list of numbers may be given weights that reflect the importance of the numbers. To calculate the weighted mean, multiply each number by its weight, add the resulting numbers and then divide by the sum of the weights. See also arithmetic mean and geometric mean.
whole number	See integer. See also fraction and decimal.

Xx

***x* coordinate**	In coordinate geometry, the first component of any point. For example, the *x* coordinate of the point (2, 3) is 2.
***x*-axis**	In coordinate geometry, the horizontal axis. 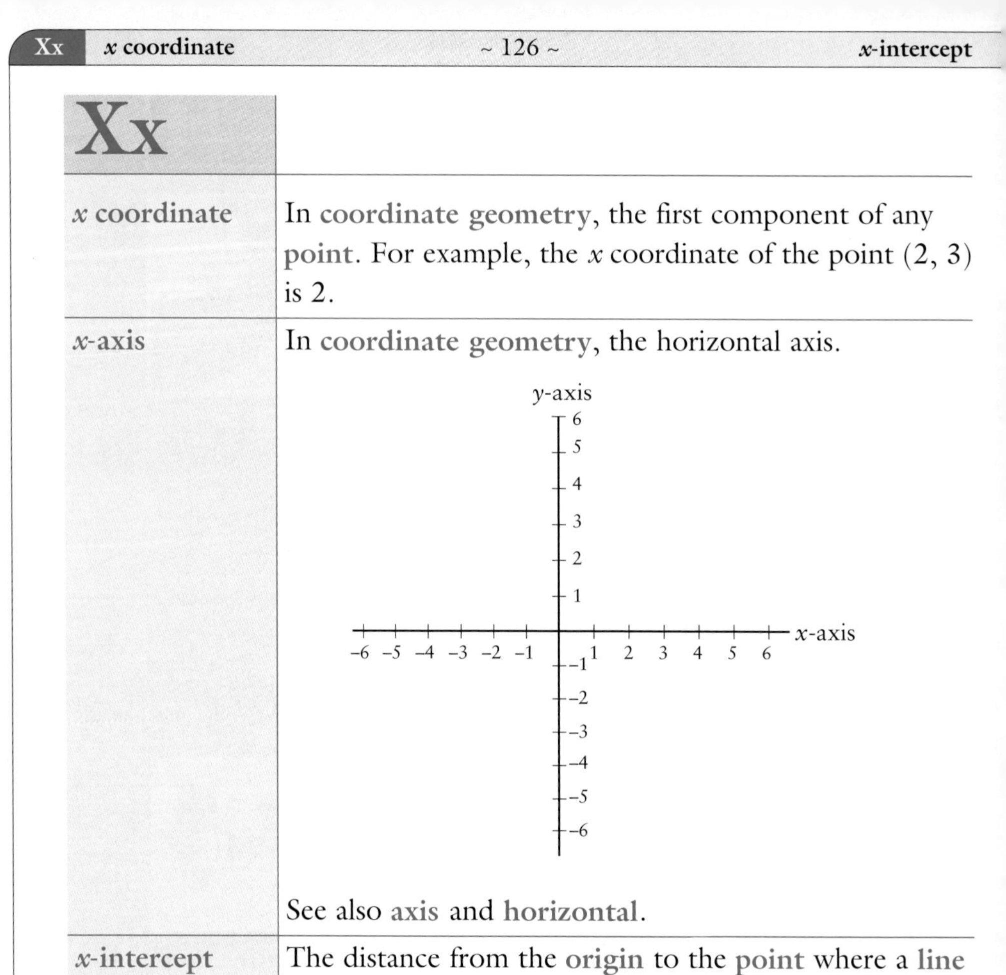 See also axis and horizontal.
***x*-intercept**	The distance from the origin to the point where a line or curve intersects the *x*-axis.

Yy

***y* coordinate**	In **coordinate geometry**, the second component of any **point**. For example, the *y* coordinate of the point (2, 3) is 3.
***y*-axis**	In **coordinate geometry**, the vertical axis. 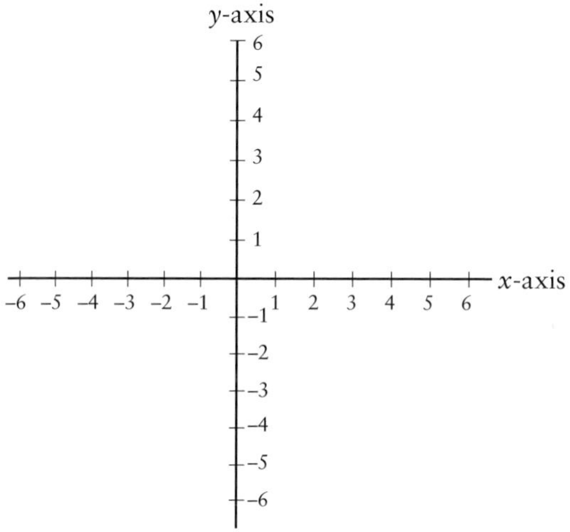 See also **axis** and **vertical**.
***y*-intercept**	The distance from the **origin** to the **point** where a **line** or curve intersects the ***y*-axis**.

Zz

zero	The digit 0.
z-score	See **standard score**.

100 Formulae and Key Symbols Explained

	Formula	The symbols represent	When is the formula used?
Algebra			
1	$x=\frac{-b\pm\sqrt{b^2-4ac}}{2a}$	a, b and c are the coefficients in the equation $ax^2+bx+c=0$	Used to solve a quadratic equation.
2	$\alpha+\beta=-\frac{b}{a}$ $\alpha\beta=\frac{c}{a}$	α and β are the roots of the quadratic equation $ax^2+bx+c=0$	Used to find the value of an expression involving α and β. Used to establish a relationship between the coefficients given a relationship between the roots.
3	$\alpha+\beta+\gamma=-\frac{b}{a}$ $\alpha\beta+\alpha\gamma+\beta\gamma=\frac{c}{a}$ $\alpha\beta\gamma=-\frac{d}{a}$	α, β and γ are the roots of the cubic equation $ax^3+bx^2+cx+d=0$	Used to find a missing coefficient in a cubic equation. Used to find a missing root of a cubic equation.
4	$x^2-(\alpha+\beta)x+\alpha\beta=0$	α and β are the roots of a quadratic equation.	Used to form a quadratic equation.
5	$x^2-y^2=(x+y)(x-y)$ $x^3-y^3=(x-y)(x^2+xy+y^2)$ $x^3+y^3=(x+y)(x^2-xy+y^2)$ $x^2-2xy+y^2=(x-y)^2$ $x^2+2xy+y^2=(x+y)^2$	x and y can be any expression or number.	Used to factorise.
6	$a^0=1$ $a^{-x}=\frac{1}{a^x}$ $a^{1/x}=\sqrt[x]{a}$ $a^{x/y}=\left(\sqrt[y]{a}\right)^x$	a is any number. x and y are numbers that appear in the index.	Used to eliminate 0, a negative number or a fraction from an index.

	Formula	The symbols represent	When is the formula used?
7	$a^x \,.a^y = a^{x+y}$ $\frac{a^x}{a^y} = a^{x-y}$ $\left(a^x\right)^y = a^{xy}$ $\left(ab\right)^u = a^x.b^x$ $\left(\frac{a}{b}\right)^x = \frac{a^x}{b^x}$	a and b are numbers that appear in the base. x and y are indices.	Used to perform calculations involving indices.
8	$\log_a 1 = 0$ $\log_a a = 1$ $\log_a a^x = x$	a is the base of a log.	Used to eliminate the word log.
9	$\log_a x + \log_a y = \log_a xy$ $\log_a x - \log_a y = \log_a \frac{x}{y}$ $p\log_a x = \log_a x^p$	a is the base of a log.	Used to convert two or more logs into one when solving a log equation. These formulae are used in reverse in differentiation to split the log of a complicated function into two or more logs of simpler functions.
10	$\log_a x = \frac{\log_b x}{\log_b a}$ $\log_a x = \frac{1}{\log_x a}$	a is the base of a log. b and x are alternative bases.	The 'change of base rule.' Used in equations when the bases are different. Also used in differentiation if the base is not e.
11	$\log_a x = \log_a y \Rightarrow x = y$ $\log_a x = b \Rightarrow a^b = x$	a is the base of a log.	Used in log equations to remove the word log.

	Formula	The symbols represent	When is the formula used?
12	If $f(x)$ is a polynomial and $f(a) = 0$, then $x - a$ is a factor of $f(x)$.	$f(x)$ is any polynomial, for example, $x^3 - 3x^2 + 2$. a is a number that makes the polynomial = 0 when substituted for x.	Used to factorise a cubic expression.
	Complex Numbers		
13	$\lvert a + bi \rvert = \sqrt{a^2 + b^2}$	a is the real part and b is the imaginary part of a complex number.	Used to find the modulus of a complex number.
14	$(\cos\theta + i\sin\theta)^n = \cos n\theta + i\sin n\theta$	$\cos\theta + i\sin\theta$ is a complex number in polar form. n is an index.	Used to raise a complex number, in polar form, to a power.
	Matrices		
15	$\dfrac{1}{ad - bc}\begin{pmatrix} d & -b \\ -c & a \end{pmatrix}$	$\begin{pmatrix} a & b \\ c & d \end{pmatrix}$ is a matrix.	Used to find the inverse of a matrix.
16	$(P^{-1}MP)^n = P^{-1}M^nP$	M is a matrix. $P^{-1}MP$ is a diagonal matrix.	Used to find M^n.
	Sequences & Series		
17	(a) $u_n = a + (n - 1)d$ (b) $S_n = \dfrac{n}{2}\left[2a + (n - 1)d\right]$ (c) $u_{n+1} - u_n = d$	a is the first term of an arithmetic sequence/series. d is the common difference. u_n is the general term of the sequence/series. u_{n+1} is the next term. n is the position of the term u_n in the sequence/series. S_n is the sum of the first n terms of an arithmetic series.	(a) Used to find a particular term of an arithmetic sequence/series. (b) Used to find the sum of the first n terms of an arithmetic series. (c) Used to find the common difference in an arithmetic sequence.

	Formula	The symbols represent	When is the formula used?
18	(a) $u_n = ar^{n-1}$ (b) $S_n = \frac{a(1-r^n)}{1-r}$, $-1 < r < 1$ (c) $S_n = \frac{a(r^n-1)}{r-1}$, $\lvert r \rvert > 1$ (d) $\frac{u_{n+1}}{u_n} = r$	a is the first term of a geometric sequence/series. r is the common ratio. u_n is the general term of the sequence/series. u_{n+1} is the next term. n is the position of the term u_n in the sequence/series. S_n is the sum of the first n terms of a geometric series.	(a) Used to find a particular term of a geometric sequence/series. (b) Used to find the sum of the first n terms of a geometric series when $\lvert r \rvert < 1$. (c) Used to find the sum of the first n terms of a geometric series when $\lvert r \rvert > 1$. (d) Used to find the common ratio of a geometric series.
19	$u_n = S_n - S_{n-1}$	u_n is the general term of a series. S_n is the sum of the first n terms of the series. S_{n-1} is the sum of the first $n - 1$ terms of the series.	Used to find the general term, u_n, of a series when we have S_n, the formula for the sum of the series.
20	$\sum_{r=1}^{n} k = kn$	n is the number of terms to be added. k is a constant.	k is a constant. When n constants are added the sum may be obtained by multiplying the constant k by n.
21	$\sum_{r=1}^{n} r = \frac{n(n+1)}{2}$	n is the number of terms to be added. S_n is the sum of the first n terms.	Used to find the sum of the first n natural numbers, excluding 0: $1 + 2 + 3 + 4 + \ldots\ldots\ldots$

	Formula	The symbols represent	When is the formula used?
22	$\sum_{r=1}^{n} r^2 = \frac{n(n+1)(2n+1)}{6}$	n is the number of terms to be added. S_n is the sum of the first n terms.	Used to find the sum of the first n terms of the series $1^2 + 2^2 + 3^2 + 4^2 + \ldots\ldots$
23	$S_\infty = \frac{a}{1-r}, -1 < r < 1$	a is the first term of an infinite geometric series. r is the common ratio. S_∞ is the sum to infinity.	Used to find the sum to infinity of an infinite, convergent geometric series.
	Binomial Theorem		
24	$(x+y)^n = \binom{n}{0}x^n + \binom{n}{1}x^{n-1}y + \binom{n}{2}x^{n-2}y^2 + \ldots\ldots + \binom{n}{n}y^n$	x and y are any two terms in a binomial expansion. n is the index.	Used to create a binomial expansion. Used in probability when a Bernoulli experiment is conducted n times. Used with de Moivre's theorem to create trigonometric identities.
25	$T_{r+1} = \binom{n}{r} x^{n-r} y^r$	x and y are any two terms in a binomial expansion. n is the index. The value of r determines which term in the expansion is represented.	The general term in a binomial expansion. Used to select a particular term without listing all the other terms.

	Formula	The symbols represent	When is the formula used?
26	$\binom{n}{r} = \dfrac{n!}{r!(n-r)!}$ $= \dfrac{n(n-1)(n-2).....(n-r+1)}{r!}$	In combinations, n is the number of objects available for selection, r is the number to be selected. In a binomial expansion, n is the index. The value of r determines which term in the expansion is represented.	Used in combinations to find the number of different selections containing r objects that can be made from n different objects. Used to evaluate $\binom{n}{r}$ without a calculator.
27	$\binom{n}{r} + \binom{n}{r+1} = \binom{n+1}{r+1}$	In combinations, n is the number of objects available for selection, r is the number to be selected.	Pascal's theorem, used to create Pascal's triangle: 1 1 1 1 2 1 1 3 3 1 1 4 6 4 1 1 5 10 10 5 1 1 6 15 20 15 6 1
	Differentation		
28	$\dfrac{dy}{dx} = f'(x) = \underset{h\to 0}{Lim}\dfrac{f(x+h) - f(x)}{h}$	$f(x)$ is some function of x. $f'(x) = \dfrac{dy}{dx}$ is the derivative. h is the horizontal distance between two points on a curve.	Used to differentiate from first principles.
29	$y = [f(x)]^n \Rightarrow$ $\dfrac{dy}{dx} = n[f(x)]^{n-1}.f'(x)$	$f(x)$ is some function of x. $f'(x) = \dfrac{dy}{dx}$ is the derivative.	The chain rule, used to differentiate a function to a power.

	Formula	The symbols represent	When is the formula used?
30	$y = \sin^{-1} f(x) \Rightarrow$ $\dfrac{dy}{dx} = \dfrac{f'(x)}{\sqrt{1 - f(x)^2}}$	$f(x)$ is some function of x. $f'(x) = \dfrac{dy}{dx}$ is the derivative.	Used to differentiate $\sin^{-1} f(x)$. The maths tables contain the less general rule $y = \sin^{-1} \dfrac{x}{a}$ $\dfrac{dy}{dx} = \dfrac{1}{\sqrt{a^2 - x^2}}$
31	$y = \tan^{-1} f(x) \Rightarrow$ $\dfrac{dy}{dx} = \dfrac{f'(x)}{1 + f(x)^2}$	$f(x)$ is some function of x. $f'(x) = \dfrac{dy}{dx}$ is the derivative.	Used to differentiate $\tan^{-1} f(x)$. The maths tables contain the less general rule $y = \tan^{-1} \dfrac{x}{a}$ $\dfrac{dy}{dx} = \dfrac{1}{a^2 + x^2}$
32	$y = \log_e f(x) \Rightarrow$ $\dfrac{dy}{dx} = \dfrac{f'(x)}{f(x)}$	$f(x)$ is some function of x. $f'(x) = \dfrac{dy}{dx}$ is the derivative.	Used to differentiate log functions when the base is e.
33	$x_{n+1} = x_n - \dfrac{f(x_n)}{f'(x_n)}$	$f(x) = 0$ is an equation. x_n is an approximation to a root of the equation, x_{n+1} is a better approximation. $f'(x)$ is the derivative of $f'(x)$.	The Newton-Raphson formula, used to find an approximation to a root of an equation (usually a cubic equation).

	Formula	The symbols represent	When is the formula used?
	Integration		
34	$\int \frac{f'(x)}{f(x)} dx = \log_e \lvert f(x) \rvert + C$	$f(x)$ is some function of x and $f'(x)$ is the derivative.	Used to integrate a fraction when the top is the derivative of the bottom.
35	$\int \frac{f'(x)dx}{a^2 + f(x)^2} = \frac{1}{a} \tan^{-1} \frac{f(x)}{a} + C$	$f(x)$ is some function of x and $f'(x)$ is the derivative.	Used to integrate a fraction that is in the form $\frac{f'(x)}{a^2 + f(x)^2}$
36	$\int \frac{f'(x)dx}{\sqrt{a^2 - f(x)^2}} = \sin^{-1} \frac{f(x)}{a} + C$	$f(x)$ is some function of x and $f'(x)$ is the derivative.	Used to integrate a fraction that is in the form $\frac{f'(x)}{\sqrt{a^2 - f(x)^2}}$
37	$\text{Area} = \left\lvert \int_a^b [f(x) - g(x)]dx \right\rvert$	$y = f(x)$ and $y = g(x)$ are the equations of two curves. a and b are two numbers on the x-axis.	Used to find the area between the curves $y = f(x)$ and $y = g(x)$ and the vertical lines $x = a$ and $x = b$. The curves must not intersect between a and b.
38	$\text{Area} = \left\lvert \int_a^b [f(y) - g(y)]dy \right\rvert$	$x = f(y)$ and $x = g(y)$ are the equations of two curves. a and b are two numbers on the x-axis.	Used to find the area between the curves $x = f(y)$ and $x = g(y)$ and the horizontal lines $y = a$ and $y = b$. The curves must not intersect between a and b.

	Formula	The symbols represent	When is the formula used?
39	Volume $= \pi\int_a^b [f(x)]^2 dx$	$y = f(x)$ is the equation of a curve. a and b are two numbers on the x-axis.	Used to find the volume of the solid formed by rotating the curve $y = f(x)$ about the x-axis between the vertical lines $x = a$ and $x = b$.
40	Volume $= \pi\int_a^b [f(y)]^2 dy$	$x = f(y)$ is the equation of a curve. a and b are two numbers on the y-axis.	Used to find the volume of the solid formed by rotating the curve $x = f(y)$ about the y-axis between the horizontal lines $y = a$ and $y = b$.
	The Circle		
41	$x^2 + y^2 = r^2$	r is the radius of a circle whose centre is (0,0). (x, y) is any point on the circle.	Used to find the equation of a circle whose centre is (0, 0) and radius is r.
42	$(x - h)^2 + (y - k)^2 = r^2$	r is the radius of a circle. (h, k) is the centre. (x, y) is any point on the circle.	Used to find the equation of a circle whose centre is (h, k) and radius is r, when we know the centre.
43	$x^2 + y^2 + 2gx + 2fy + c = 0$	$(-g, -f)$ is the centre of a circle and $\sqrt{g^2 + f^2 - c}$ is the radius. (x, y) is any point on the circle.	Used to find the equation of a circle when we don't know the centre. We will be given sufficient information to create three equations involving g, f and c.

	Formula	The symbols represent	When is the formula used?
44	$g^2 = c$	$(-g, -f)$ is the centre of a circle and $\sqrt{g^2 + f^2 - c}$ is the radius.	Used when we know that the x-axis is a tangent to the circle. With three equations in g, f and c, we can find their values using simultaneous equations and hence find the equation of the circle.
45	$f^2 = c$	$(-g, -f)$ is the centre of a circle and $\sqrt{g^2 + f^2 - c}$ is the radius.	Used when we know that the y-axis is a tangent to the circle. With three equations in g, f and c, we can find their values using simultaneous equations and hence find the equation of the circle.
		Vectors	
46	$\overrightarrow{ab} + \overrightarrow{bc} = \overrightarrow{ac}$	$\overrightarrow{ab}$, $\overrightarrow{bc}$ and $\overrightarrow{ac}$ are vectors.	Used to add vectors if one of them begins where the other ends.
47	$\overrightarrow{ab} = \vec{b} - \vec{a}$	$\overrightarrow{ab}$ is the vector from a to b. $\vec{a}$ and $\vec{b}$ are vectors that begin at the origin and end at a and b respectively.	Used to express a two-letter vector, $\overrightarrow{ab}$ in terms of the single-letter vectors $\vec{a}$ and $\vec{b}$.

	Formula	The symbols represent	When is the formula used?
48	$\vec{p} = \dfrac{m\vec{b} + n\vec{a}}{m + n}$	$\vec{a}$, $\vec{b}$ and $\vec{p}$ are vectors that begin at the origin. The point p divides the line segment $[ab]$ internally in the ratio $m{:}n$.	Used to express the vector $\vec{p}$ in terms of the vectors $\vec{a}$ and $\vec{b}$ when the point p divides the line segment $[ab]$ internally in the ratio $m{:}n$.
49	$\vec{p} = \frac{1}{3}\vec{a} + \frac{1}{3}\vec{b} + \frac{1}{3}\vec{c}$	a, b and c are the vertices of a triangle. p is the centroid.	Used to express the vector from the origin to the centroid of a triangle, in terms of the vectors from the origin to each of the vertices.
50	$\vec{r} = a\vec{i} + b\vec{j}$ $\vec{r}^{\perp} = -b\vec{i} + a\vec{j}$	$\vec{r} = a\vec{i} + b\vec{j}$ is any vector. $\vec{r}^{\perp} = -b\vec{i} + a\vec{j}$ is the related perpendicular vector.	Used to create the related perpendicular vector, given a vector expressed in terms of $\vec{i}$ and $\vec{j}$.
51	$\lvert a\vec{i} + b\vec{j}\rvert = \sqrt{a^2 + b^2}$	$a\vec{i} + b\vec{j}$ is any vector, expressed in terms of $\vec{i} + \vec{j}$. $\lvert a\vec{i} + b\vec{j}\rvert$ is the norm or length of the vector.	Used to calculate the length of a vector when the vector is expressed in terms of $\vec{i}$ and $\vec{j}$.
52	$\dfrac{\vec{x}}{\lvert\vec{x}\rvert}$	$\vec{x}$ is any vector. $\lvert\vec{x}\rvert$ is the length of the vector.	Used to find the unit vector in the direction of a given vector.
53	$(a\vec{i} + b\vec{j}).(c\vec{i} + d\vec{j})$ $= ac + bd$	$a\vec{i} + b\vec{j}$, $c\vec{i} + d\vec{j}$ and vectors.	Used to find the dot product of two vectors when they are written in terms of $\vec{i}$ and $\vec{j}$.

	Formula	The symbols represent	When is the formula used?
54	$\dfrac{a\vec{i} + b\vec{j}}{\sqrt{a^2 + b^2}}$	$a\vec{i} + b\vec{j}$ is any vector, expressed in terms of $\vec{i}$ and $\vec{j}$. $a^2 + b^2$ is the norm or length of the vector	Used to create the unit vector in the direction of $a\vec{i} + b\vec{j}$.
55	$\vec{p}.\vec{q} = \lvert\vec{p}\rvert.\lvert\vec{q}\rvert.\cos\theta$	$\vec{p}$ and $\vec{q}$ are vectors that begin at the origin. θ is the angle between the vectors.	Used to calculate the scalar (dot) product of $\vec{p}$ and $\vec{q}$.
56	$\cos\theta = \dfrac{\vec{p}.\vec{q}}{\lvert\vec{p}\rvert.\lvert\vec{q}\rvert}$	$\vec{p}$ and $\vec{q}$ are vectors that begin at the origin. $\lvert\vec{p}\rvert$, $\lvert\vec{q}\rvert$ are the lengths of $\vec{p}$ and $\vec{q}$, respectively. θ is the angle between the vectors $\vec{p}$ and $\vec{q}$.	Used to find the measure of θ, the angle between the vectors $\vec{p}$ and $\vec{q}$.
	The Line		
57	$\sqrt{(x_2 - x_1)^2 + (y_2 - y_1)^2}$	(x_1, y_1) and (x_2, y_2) are two points.	Used to calculate the distance between the two points.
58	$\left(\dfrac{x_1 + x_2}{2}, \dfrac{y_1 + y_2}{2}\right)$	(x_1, y_1) and (x_2, y_2) are two points.	Used to calculate the midpoint between (x_1, y_1) and (x_2, y_2).
59	$p\left(\dfrac{mx_2 + nx_1}{m + n}, \dfrac{my_2 + my_1}{m + n}\right)$	$a(x_1, y_1)$ and $b(x_2, y_2)$ are two points. The point p divides the line segment $[ab]$ internally in the ratio $m{:}n$.	Used to find the coordinates of the point p that divides the line segment $[ab]$ internally in the ratio $m{:}n$.
60	$\left(\dfrac{mx_2 - nx_1}{m - n}, \dfrac{my_2 - ny_1}{m - n}\right)$	$a(x_1, y_1)$ and $b(x_2, y_2)$ are two points. The point p divides the line segment $[ab]$ externally in the ratio $m{:}n$.	Used to find the coordinates of the point p that divides the line segment $[ab]$ externally in the ratio $m{:}n$.

	Formula	The symbols represent	When is the formula used?
61	$m = \dfrac{y_2 - y_1}{x_2 - x_1}$	$a(x_1, y_1)$ and $b(x_2, y_2)$ are two points. m is the slope of the line that passes through a and b.	Used to find the slope of a line when we know two points on the line.
62	$y - y_1 = m(x - x_1)$	(x_1, y_1) is a point on a line. m is the slope of the line.	Used to find the equation of a line when we know a point on it and the slope.
63	$\text{Area} = \dfrac{1}{2}\lvert x_1y_2 - x_2y_1 \rvert$	$a(x_1, y_1)$ and $b(x_2, y_2)$ are two points.	Used to find the area of a triangle with vertices (x_1, y_1), (x_2, y_2) and $(0, 0)$.
64	$d = \left\lvert \dfrac{ax_1 + by_1 + c}{\sqrt{a^2 + b^2}} \right\rvert$	$ax + by + c = 0$ is the equation of a line. (x_1, y_1) is a point not on the line. d is the perpendicular distance from the point to the line.	Used to calculate the perpendicular distance from a point to a line. Also used to find a missing coefficient in the equation of a line when we know the distance from some point to the line.
65	$\tan \theta = \pm \dfrac{m_1 - m_2}{1 + m_1m_2}$	m_1 and m_2 are the slopes of two lines and θ is the angle between the lines.	Used to find the measure of the angle between two lines when we know the slopes of the lines. Also used to find one of the slopes when we know the angle.
66	$g = \left(\dfrac{x_1 + x_2 + x_3}{3}, \dfrac{y_1 + y_2 + y_3}{3} \right)$	(x_1, y_1), (x_2, y_2) and (x_3, y_3) are the vertices of a triangle. g is the centroid.	Used to find the coordinates of the centroid of a triangle when we know the vertices.

	Formula	The symbols represent	When is the formula used?
67	$\mu(a_1x + b_1y + c_1) + \lambda(a_2x + b_2y + c_2) = 0$	$a_1x + b_1y + c_1 = 0$ is the equation of a line. $a_2x + b_2y + c_2 = 0$ is the equation of a second line. μ and λ are numbers.	Used to find the equation of a line that passes through the point of intersection of two given lines without finding the coordinates of the point of intersection.
68	$\dfrac{a_1x + b_1y + c_1}{\sqrt{a_1^2 + b_1^2}} = \pm \dfrac{a_2x + b_2y + c_2}{\sqrt{a_2^2 + b_2^2}}$	$a_1x + b_1y + c_1 = 0$ is the equation of a line. $a_2x + b_2y + c_2 = 0$ is the equation of a second line.	Used to find the equations of the bisectors of the angles between two lines, given the equations of the lines.
69	$x = x_1 + (x_2 - x_1)t$ $y = y_1 + (y_2 - y_1)t, \quad t \in R$	$a(x_1, y_1)$ and $b(x_2, y_2)$ are two points. (x, y) represents any point on the line ab. t can be any real number.	Used to find parametric equations of the line ab, given the coordinates of a and b.
70	$x = x_1 + (x_2 - x_1)t$ $y = y_1 + (y_2 - y_1)t, \quad 0 \leq t \leq 1$	$a(x_1, y_1)$ and $b(x_2, y_2)$ are two points. (x, y) represents any point on the line segment $[ab]$. t can be any real number between 0 and 1, inclusive.	Used to find parametric equations of the line segment $[ab]$, given the coordinates of a and b.

	Formula	The symbols represent	When is the formula used?
	Trigonometry		
71	$\lim_{\theta \to 0} \frac{\sin\theta}{\theta} = 1$	θ is an angle, measured in radians.	Used to evaluate trigonometric limits and to differentiate trigonometric functions from first principles.
72	$a^2 = b^2 + c^2 - 2bc\cos A$	a, b and c are the lengths of the sides in a triangle. A is the measure of the angle between sides b and c.	The cosine rule, used to find a side of a triangle when we know the lengths of the other two sides and the measure of the angle between them.
73	$\cos A = \frac{b^2 + c^2 - a^2}{2bc}$	a, b and c are the lengths of the sides in a triangle. A is the measure of the angle between sides b and c.	A rearrangement of the cosine rule, used to find the measure of an angle in a triangle when we know the lengths of all the sides.
74	$\frac{\sin A}{a} = \frac{\sin B}{b} = \frac{\sin C}{c}$	a, b and c are the lengths of the sides of a triangle. A, B and C are the measures of their respective opposite angles.	The sine rule, used to find the measure of an angle or the length of a side in a triangle when we know a side, its opposite angle and another side or angle.
75	$\cos^2 A + \sin^2 A = 1$ $\cos^2 A - \sin^2 A = \cos 2A$ $2\sin A \cos A = \sin 2A$ $\cos 2A = 1 - 2\sin^2 A$ $\cos 2A = 2\cos^2 A - 1$	A is any angle. $2A$ is twice the angle A.	Important trigonometric formulae used to prove identities and solve equations.

	Formula	The symbols represent	When is the formula used?
Permutations & Combinations			
76	$n!$	n is the number of objects to be arranged.	In permutations, used to find the number of arrangements of n different objects when all n objects are to appear in each arrangement.
77	$\frac{n!}{p!q!}$	n is the number of objects to be arranged. There are p identical objects of one type and q identical objects of a different type.	In permutations, used to find the number of arrangements of n objects when some of them are identical.
78	${}^{n}P_{r} = \frac{n!}{(n-r)!}$	n is the number of objects to be arranged and r is the number of objects to appear in each arrangement.	In permutations, used to find the number of arrangements of n different objects using r objects in each arrangement.
79	$m!(n-m+1)!$	n is the number of objects to be arranged and m is the number of objects that must remain together.	The number of arrangements of n different objects when m of those objects must remain together.
80	$\binom{n}{r} = \frac{n!}{r!(n-r)!}$	n is the number of objects available for selection and r is the number to be selected.	In combinations, used to find the number of selections containing r objects that can be made from a set containing n different objects.

	Formula	The symbols represent	When is the formula used?
81	$\binom{x}{a} \times \binom{y}{b}$	x is the number of objects available for selection in one category, and a is the number of those to be selected. y is the number of objects available for selection in a different category, and b is the number of those to be selected.	In combinations, to find the number of selections which can be made when some of them must come from one category and the others must come from a different category.
Statistics & Difference Equations			
82	$\bar{x} = \frac{\sum xf}{\sum f}$	$\bar{x}$ is the mean. x is a number and f is the frequency of that number.	Used to calculate the mean of a frequency distribution.
83	$\sigma = \sqrt{\frac{\sum_{i=1}^{n} f(x_i - \bar{x})^2}{\sum_{i=1}^{n} f_i}}$	σ is standard deviation. x_1 is an individual statistic. f_1 is its frequency. $\bar{x}$ is the mean. $x_1 - \bar{x}$ is the deviation of a statistic. n is the number of figures involved.	Used to calculate the standard deviation of a frequency distribution.
84	$u_n = l\alpha^n + m\beta^n$	u_n is any term in a sequence. α and β are the roots of the characteristic equation. l and m are obtained from the initial conditions.	Used to solve a difference equation.

	Formula	The symbols represent	When is the formula used?
	Probability		
85	$P(E) = \dfrac{\#(E)}{\#(\text{Sample Space})}$	E is some event resulting from an experiment in probability. P(E) is the probability that E will occur. #(Sample Space) is the total number of possible outcomes. #(E) is the number of outcomes in the category E.	Used to calculate the probability that the event E will occur when an experiment takes place in probability.
86	$P(\text{not } E) = 1 - P(E)$	E is some event resulting from an experiment in probability. P(E) is the probability that E will occur. P(not E) is the probability that E will not occur.	Used to calculate the probability that the event E will not occur when an experiment takes place in probability.
87	$P(E\|F) = \dfrac{\#(E \cap F)}{\#F}$	E and F are two events resulting from an experiment in probability.	Used to calculate the probability that E will occur, given that F has already occurred.
88	$P(E\|F) = P(E)$	E and F are two events resulting from an experiment in probability.	Used to prove that E and F are independent events.
89	$P(E\|F) \neq P(E)$	E and F are two events resulting from an experiment in probability.	Used to prove that E and F are dependent events.

	Formula	The symbols represent	When is the formula used?
90	$P(E \cap F) = P(E).P(F)$	E and F are two independent events resulting from an experiment in probability.	When successive experiments take place, the formula may be applied to calculate the probability that E occurs first, then F. It can also be applied to calculate the probability that in a single experiment the outcome is both E and F.
91	$P(E \cap F) = P(E).P(F\|E)$	E and F are two dependent events resulting from an experiment in probability.	When successive experiments take place, the formula may be applied to calculate the probability that E occurs first, then F. It can also be applied to calculate the probability that in a single experiment the outcome is both E and F.
92	$P(E \cup F) = P(E) + P(F)$	E and F are two mutually exclusive events resulting from an experiment in probability.	In a single experiment, used to calculate the probability that E or F will occur when E and F are mutually exclusive.
93	$P(E \cup F) = P(E) + P(F) - P(E \cap F)$	E and F are two events that are not mutually exclusive, resulting from an experiment in probability.	In a single experiment, used to calculate the probability that E or F will occur when E and F are not mutually exclusive.

	Formula	The symbols represent	When is the formula used?
	Probability Option		
94	$z = \dfrac{x - \bar{x}}{}$	x is the outcome of an experiment in probability which is to be converted to standard units (z). $\bar{x}$ is the mean. σ is the standard deviation.	Used to convert the result of an experiment to standard units.
95	$\bar{x}_1 - 1.96\sigma_{\bar{x}} \leq \mu \leq \bar{x}_1 + 1.96\sigma_{\bar{x}}$	$\bar{x}_1$ is the mean of a sample. μ is the population mean. $\sigma_{\bar{x}}$ is the standard error.	Used to calculate the 95% confidence interval for the mean.
96	$\sigma_{\bar{x}} = \dfrac{\sigma}{\sqrt{n}}$	$\sigma_{\bar{x}}$ is the standard error. σ is the standard deviation. n is the size of the sample.	Used to calculate the standard error.
97	$z_1 = \dfrac{\bar{x}_1 - \mu_{\bar{x}}}{\sigma_{\bar{x}}}$	$\bar{x}_1$ is the mean of a sample. $\mu_{\bar{x}} = \mu$ is the population mean. $\sigma_{\bar{x}}$ is the standard error. z_1 is the number of standard units.	Used to identify how many standard units, above or below the true mean, a given sample mean is.

	Formula	The symbols represent	When is the formula used?
	Further Calculus Option		
98	If $\left\lvert \lim_{n\to\infty} \frac{u_{n+1}}{u_n} \right\rvert < 1$ $\sum^{\infty} u_n$ is a convergent series	u_n is any term in a series and u_{n+1} is the next term.	The ratio test, used to determine whether an infinite series is convergent or divergent.
99	$f(x) = \frac{f(0)}{0!} + \frac{f'(0)}{1!}x + \frac{f''(0)}{2!}x^2 + \frac{f'''(0)}{3!}x^3 + \ldots$	$f(x)$ is some function of x. $f'(x)$ is the first derivative. $f''(x)$ is the second derivative, etc.	Used to create a Maclaurin series.
100	$1 - x^2 + x^4 - x^6 + x^8 \ldots\ldots\ldots$	The Maclaurin series for $\frac{1}{1 + x^2}$	Used to create the Maclaurin series for $\tan^{-1} x$.

Symbol	Reference
$+$	Plus: to add two numbers.
$-$	Minus: to subtract two numbers.
$\times$	Multiplication: to multiply two numbers.
$\div$	Division: to divide two numbers.
$\pm$	Plus or minus.
$\sqrt{\ }$	Square root.
$\sqrt[n]{\ }$	n^{th} root.
N	Natural numbers.
N_0	Natural numbers excluding zero.
Z	Integers.
Q	Rational numbers.
R	Real numbers.
C	Complex numbers.
$=$	Equals.
$\approx$	Is approximately equal to.
$\neq$	Is not equal to.
$\equiv$	Is congruent to.
$>$	Is bigger than.
$<$	Is less than.
$\geq$	Is bigger than or equal to.
$\leq$	Is less than or equal to.
$-2 \leq x \leq 3, x \in R$	Every real number between −2 and 3, inclusive.
$-2 < x < 3, x \in R$	Every real number between −2 and 3, excluding −2 and 3.
$-2 \leq x \leq 3, x \in Z$	Every whole number between −2 and 3, inclusive.

Symbol	Reference
$-2 < x < 3, x \in Z$	Every whole number between –2 and 3, excluding –2 and 3.
%	Per cent (per hundred).
$\{1, 2, 3\}$	A set containing the elements 1, 2 and 3.
$\{\ \}$ or $\varnothing$	The empty set.
#	Cardinal number.
$\in$	Is an element of.
$\notin$	Is not an element of.
$\cap$	Intersection.
$\cup$	Union.
$\setminus$	Not.
A'	The complement of the set A.
∞	Infinity.
π	Pi. Greek letter (lowercase). The number obtained when the circumference of a circle is divided by the diameter. $\pi \approx 3.14$.
θ	Theta. Greek letter (lowercase). Usually used to represent an unknown angle.
α	Alpha. Greek letter (lowercase). Often used to represent a root of a quadratic equation.
β	Beta. Greek letter (lowercase). Often used to represent a root of a quadratic equation.
ω	Omega. Greek letter (lowercase). Often used to represent a cube root of 1.
λ	Lamda. Greek letter (lowercase).
μ	Mu. Greek letter (lowercase).
σ	Sigma. Greek letter (lowercase). Usually used to represent standard deviation in statistics.

Symbol	Reference
$\sigma_{\bar{x}}$	Standard error of the mean. (Probability option).
Σ	Sigma. Greek letter (capital). Used as a short way to represent a sum.
$\sum_{r=1}^{5} r^2$	$1^2 + 2^2 + 3^2 + 4^2 + 5^2$
e	The exponential constant. $e = \lim_{n\to\infty} \left(\frac{n+1}{n}\right)^n \approx 2.718$ Or In groups, e is used to represent the identity element of a group.
i	In complex numbers, $i = \sqrt{-1}$.
a, b, c, etc.	In geometry, symbols to represent points on the plane. In algebra, symbols for unknown numbers.
ab	In geometry, an infinite line, passing through the points a and b. In algebra, the number a multiplied by the number b.
$[ab]$	A line segment, from the point a to the point b.
$\lvert ab \rvert$	The distance from a to b.
$\lvert x \rvert$	The modulus or absolute value of x. It is the distance from x to zero on a number line.
$\lvert z \rvert$	In complex numbers, $\lvert z \rvert$ is the distance from z to zero. If $z = a + bi$ then $\lvert z \rvert = \sqrt{a^2 + b^2}$.
$\perp$	Perpendicular to.
$\angle abc$	The angle formed by joining a to b and b to c.
$\lvert \angle abc \rvert$	The size or measure of the angle abc.
$\overrightarrow{ab}$	In transformation geometry, the translation $\overrightarrow{ab}$. In vectors, a vector from the point a to the point b.
$\overrightarrow{a}$	The vector from the origin to the point a.
S_a	Central symmetry in the point a.
S_o	Central symmetry in the origin.

Symbol	Reference
S_{ab}	Axial symmetry in the line *ab*.
S_X	Axial symmetry in the *x*-axis.
S_Y	Axial symmetry in the *y*-axis.
$f(x)$	Some function whose value depends on the value of the number *x*.
$f(2)$	The number obtained when *x* is replaced by 2, in the function $f(x)$.
$\overline{x}$	In statistics, the mean.
$\lim\limits_{x \to \infty} f(x)$	Evaluate the limit of the function $f(x)$ as *x* gets infinitely bigger.
$\lim\limits_{x \to 2} f(x)$	Evaluate the limit of $f(x)$ as *x* gets closer and closer to 2.
$\frac{dy}{dx}$	The derivative of *y* with respect to *x*.
$\frac{d^2y}{dx^2}$	The derivative of $\frac{dy}{dx}$ with respect to *x*, called the second derivative.
$\int f(x)dx$	The indefinite integral of the function $f(x)$.
$\int_a^b f(x)dx$	A definite integral. If $\int f(x)dx = F(x)$ then $\int_a^b f(x)dx = F(b) - F(a)$.
$n!$	Factorial *n*. $n! = n(n-1)(n-2) \ldots\ldots (1)$. Note: $0! = 1$.
$\binom{n}{r}$	$\binom{n}{r} = \frac{n!}{r!(n-r)!}$
^nP_r	$^nP_r = \frac{n!}{(n-r)!}$
$\vec{i}$	A vector from (0, 0) to (1, 0).
$\vec{j}$	A vector from (0, 0) to (0, 1).
$\vec{a}.\vec{b}$	In vectors, $\vec{a}.\vec{b} = \lvert\vec{a}\rvert \times \lvert\vec{b}\rvert \times \cos\theta$, where θ is the angle between $\vec{a}$ and $\vec{b}$.

Symbol	Reference
$\vec{r}^{\perp}$	The related perpendicular vector. If $\vec{r} = a\vec{i} + b\vec{j}$ then $\vec{r}^{\perp} = -b\vec{i} + a\vec{j}$.
$\bar{z}$	In complex numbers, if $z = a + bi$, then the complex conjugate $\bar{z} = a - bi$.
(x', y')	In coordinate geometry, the image of a point under a linear transformation.
u_n	A single term in a sequence or a series.
u_{n+1}	The term that comes after u_n.
S_n	The sum of the first n terms of a series. Or In groups, S_n represents the set of all permutations of n objects.
S_∞ or $\sum_{n=1}^{\infty} u_n$	The sum to infinity of a series.
$*$	Used in groups to represent a binary operation.
Z_n	In groups, the set of possible remainders on division by n.
I	In matrices, the identity matrix $I = \begin{pmatrix} 1 & 0 \\ 0 & 1 \end{pmatrix}$.
M^{-1}	The inverse of the matrix M.
a^{-1}	In groups, the inverse of the element a.

Notes